Riparian Ecosystem Recovery
in Arid Lands

Mark K. Briggs

Riparian Ecosystem Recovery

in Arid Lands

Strategies and References

The University of Arizona Press

Tucson

The University of Arizona Press
© 1996 The Arizona Board of Regents
All rights reserved

Library of Congress Cataloging-in-Publication Data
Briggs, Mark K. (Mark Kendig), 1961–
Riparian ecosystem recovery in arid lands : strategies and references / Mark K.
Briggs.
    p. cm.
Includes bibliographical references and index.
ISBN 0-8165-1642-1 (cloth). — ISBN 0-8165-1644-8 (paper)
1. Riparian ecology—Southwest, New. 2. Riparian ecology—Mexico.
3. Restoration ecology—Southwest, New. 4. Restoration ecology—Mexico.
5. Riparian ecology—Southwest, New—Case studies. 6. Riparian ecology—
Mexico—Case studies. 7. Restoration ecology—Southwest, New—Case studies.
8. Restoration ecology—Mexico—Case studies. I. Title.
QH104.5.S6B77   1996
333.91'8—dc20   96-9957

Manufactured in the United States of America on acid-free, archival-quality paper.

12 11 10 09 08 07   7 6 5 4 3 2

To my parents, and my friends

Kai Nyandemo and Mohammed Din

Contents

Figures and Tables

Tables

Preface

In the not too distant past, lower Rincon Creek (a portion of a small stream in the Sonoran Desert) was dominated by a riparian forest that in some areas extended more than half a kilometer from each side of the stream channel. The predominant woody vegetation was Arizona walnut *(Juglans major)*, with some of the larger individuals exceeding 5 m in circumference and towering close to 20 m above the floodplain. During this time, groundwater levels probably never dropped more than a few meters below the soil surface, and streamflow was nearly perennial. The well-being of a plethora of wildlife species depended on this riparian area. Water-loving species like the beaver *(Castor canadensis)* and the southwestern Woodhouse toad *(Bufo woodhousei australis)* lived year-round in or very near this ecosystem. The mountain lion *(Felis concolor),* coati *(Nasua nasua),* mule deer *(Odocoileus hemionus),* Yuma myotis *(Myotis yumanensis),* and vermilion flycatcher *(Pyrocephalus rubinus)* were continual visitors.

This riparian ecosystem no longer exists. During the past 60 years, the combined impacts of groundwater pumping, stream channelization, agriculture, livestock grazing, and, currently, urbanization have almost completely destroyed this area. Today, streamflow occurs only following heavy rains, and scattered remnants are all that remain of the once immense riparian forest.

Unfortunately, the devastation that befell the riparian ecosystem of lower Rincon Creek is not unique. Riparian ecosystems have disappeared at an alarming rate all across northern Mexico and the southwestern United States. Although the extent of damage is still not well understood, most observers will agree that the deterioration of riparian ecosystems in this part of the world has reached significant proportions.

This book is designed as a manual for those in the natural resource field who have taken on the task of repairing damaged riparian ecosystems. The emphasis is on evaluating riparian ecosystems so that the causes of ecological deterioration can be understood; restoration based on a sound understanding of the causes of decline will have a greater chance of succeeding. This guidebook provides readers with examples of past riparian recovery projects and the results of numerous studies on riparian ecosystems. I hope that this information will assist efforts across the Southwest and elsewhere to improve the condition of damaged riparian areas.

Acknowledgments

I feel rather awkward about being listed as the sole author of this book, because the research and writing of it were remarkable experiences involving a diversity of individuals and organizations. This project was truly a partnership, and I am honored to have had the privilege and opportunity to meet, work with, and learn from its many contributors. I want to express my appreciation to the Arizona Game and Fish Department Heritage Fund, National Fish and Wildlife Foundation, School of Renewable Natural Resources of the University of Arizona, U.S. Bureau of Land Management, U.S. Bureau of Reclamation, U.S. Fish and Wildlife Service, and World Wildlife Fund for funding the research and development of this guidebook. My heartfelt thanks to Waite Osterkamp (U.S. Geological Survey), Guy McPherson (University of Arizona), and Mary Schmid (Rincon Institute) for the hours they spent poring over the manuscript and grooming my often chaotic sentences into something readable. Many thanks also to my office mates for their patience and support over the years, and to Laura Jackson (University of Northern Iowa); Juliet Stromberg (Arizona State University); Bruce Roundy (Brigham Young University); Rick Paradis (University of Vermont); Frank Gregg, Phil Guertin, Roy Keys, Phil Ogden, William Shaw, and Erv Zube (University of Arizona); Steve Moore (Center for Image Processing in Education); Reginald Briggs (Geomega); Luther Propst (Rincon Institute); Douglas Morris and Bill Paleck (National Park Service); Leonard DeBano, Pattie Fenner, and Greg Goodwin (U.S. Forest Service); Robert Hall and Al Bammann (U.S. Bureau of Land Management); and James Roelle and Michael Scott (U.S. Fish and Wildlife Service) for reviewing, editing, and supporting this project.

Riparian Ecosystem Recovery
in Arid Lands

1/ An Overview

Riparian ecosystems are the vegetation, habitats, and ecosystems associated with bodies of water or dependent on the existence of perennial, ephemeral, or intermittent surface or subsurface drainage (Arizona Riparian Council 1996). The riparian ecosystems of the southwestern United States are some of the most productive ecosystems in North America (Johnson and Jones 1977; Johnson and McCormick 1978). In addition, they indirectly affect the stability and quality of surrounding ecosystems by reducing flood peaks, acting as sediment and nutrient sinks, controlling water temperature, and increasing groundwater recharge (Schmidt 1987). Despite their relatively small expanse (riparian ecosystems associated with perennial waters in Arizona constitute less than 1 percent of the total land area of the state, for example), riparian areas play a critical role in the life cycles of an inordinate number of wildlife species and provide important recreation opportunities for outdoor enthusiasts.

Riparian ecosystems are declining throughout the southwestern United States, and many have disappeared completely. The rapid decline of these valuable ecosystems has made riparian conservation a focal issue in the public eye, as well as for many federal, state, and private organizations. Nevertheless, progress toward checking their decline has been marginal. This is due, in part, to the fact that the science of repairing damaged riparian ecosystems is relatively young; we are still investigating fundamental questions on riparian ecosystem processes and the impacts of human activities. In addition, we have learned little from past riparian recovery efforts because the results of only a relatively small number of projects to improve the condition of damaged riparian areas have been evaluated for the benefit of future projects.

Current research findings serve as the foundation for this guidebook, which is designed to help ecosystem managers evaluate degraded riparian ecosystems so that effective strategies for improving their condition can be developed. The practical nature of this book begins to fill the gap in riparian conservation literature between the academic world and practitioners in the natural resource field. Although a technical document, the book is written so that those who do not have a background in natural resources—including students, educators, developers, public officials, and landowners—can readily use it as well.

Because this book is based for the most part on an evaluation of riparian revegetation efforts in Arizona (Briggs 1992), a principal focus is on evaluating the damaged riparian site to determine the potential effectiveness of revegetation. However, other types of recovery strategies (e.g., improved livestock grazing, in-stream modifications) are discussed as well. Although most of the lessons presented here have been learned from experiences in arid wildland areas, many of the strategies are also relevant to nonarid climates and urban areas.

Background on Riparian Ecosystems

Riparian ecosystems are characterized by high diversity in both plant and wildlife species. The mesic nature of riparian areas permits the establishment and growth of many plant species not found on adjacent, more xeric uplands (Warren and Anderson 1985). Youngblood et al. (1985) found more than 600 plant species within 469 fifty-square-meter sample plots in eastern Idaho and western Wyoming. Spear and Mullins (1987) observed that waterfowl, wintering bald eagles *(Haleaeetus leucocephalus)*, peregrine falcons *(Falco peregrinus)*, and whooping cranes *(Grus americana)* are constant visitors to the riparian ecosystem along the Rio Grande River in New Mexico. In addition, riparian ecosystems are home to many species of amphibians and reptiles. Mammals such as desert shrews *(Notiosorex crawfordi)*, hoary bats *(Lasiurus cinereus)*, mice *(Peromyscus* spp.), bobcats *(Felis rufus)*, and mule deer *(Odocoileus hemionus)* also are very much a part of these ecosystems. Hubbard (1977) described riparian areas as having a fairly high endemism including a relatively large number of endangered species.

Riparian ecosystems in arid parts of the world differ in many respects from those in more humid climates. One of the more obvious differences is the abrupt transition between the mesic riparian zone and surrounding arid areas. In the Sonoran Desert, for example, it is common to find riparian areas where you can reach out and touch a phreatophytic species such as a Fremont cottonwood tree *(Populus fremontii)* with one hand and a xerophytic species such as a saguaro cactus *(Carnegiea gigantea)* with the other.

Riparian ecosystems take on many forms and are characterized by a variety of plant communities. Riparian ecosystems can be narrow, with abrupt transitions between the riparian and upland plant communities, or broad, with the riparian zone extending for hundreds of meters from the stream channel. Change in elevation (with its concomitant effects on frequency of inundation) appears to be the most significant factor associated with the distribution of riparian plant communities and their species composition (Szaro 1989). In the Southwest, low-lying (less than 1,000 m in elevation) riparian plant communities are often dominated by velvet mesquite *(Prosopis velutina)* or Goodding willow *(Salix gooddingii)*. Mid-elevation (less than 1,500 m) riparian communities are often dominated by species such as sacaton *(Sporobolus wrightii)* or Arizona sycamore *(Platanus wrightii)*, while higher-elevation communities may be dominated by blue spruce *(Picea pungens)*.

Within an elevation range, other factors contribute strongly to determining the composition of the riparian plant community, including fluvial geomorphic processes (Osterkamp 1978; McBride and Strahan 1984), with elevation above the stream channel being a major cofactor (Nixon et al. 1977; Hupp 1982; Hupp and Osterkamp 1985). Stream bearing, stream gradient, flow regime, and geology also affect the characteristics of riparian plant communities at a local level (Zimmerman 1969; Szaro 1989).

Human-related activities have contributed to the decline of these valuable ecosystems and their associated rivers. Many of the cienegas, mesquite bosques, and forests of cottonwoods and willows that once dominated portions of southeastern Arizona have fallen under the weight of impacts from agriculture, groundwater pumping, introduced exotics, livestock grazing, fuelwood harvesting, and flood control measures, among others (Bahre 1991). Beginning in the 1860s, cottonwood and willow forests along the lower Colorado River were extensively cut to provide fuel for steamer travel (Ohmart et al. 1977). With the improvement of pumping technology in the twentieth century, the pumping of groundwater to meet the thirst of our major urban centers has led to groundwater decline, which, in turn, has contributed to the deterioration of riparian ecosystems in many areas, including the disappearance of mesquite bosques along the Santa Cruz River near Tucson and the Gila River near the Casa Grande National Monument (Betancourt and Turner n.d.). The construction of dams has led to the demise of stands of cottonwoods along the lower Salt River in central Arizona (Fenner et al. 1985) and the lower Verde River (McNatt et al. 1980) near Phoenix, Arizona (fig. 1.1). The creation of in-stream reservoirs, the diversion of water for cropland irrigation and electrical production, the demands of an increasing human population, and a variety of other human-related impacts have greatly altered the flow regime and channel dimensions of portions of the Platte River system in Colorado, Wyoming, and Nebraska (Minnich 1978). Overuse by livestock has also

Figure 1.1 This stand of cottonwoods along the lower Verde River just north of Phoenix, Arizona, met its demise as a result of the artificial manipulation of natural flow patterns from upstream dams. (Photograph by Mark Briggs)

contributed to the decline of riparian areas throughout the western United States: examples include Rodero Creek, Nevada (Swanson et al. 1987); Henry's Fork, Idaho (Platts et al. 1989); Tonto Creek, Arizona (Alford 1993); and Big Creek, Utah (Platts and Nelson 1989).

In New Mexico, roughly 21,000 ha (54,000 acres) of riparian vegetation was removed between 1967 and 1971 in the flawed belief that removal of streamside plants would increase the amount of water supplied to downstream users (Ohmart and Anderson 1986). Dense stands of nonnative saltcedar *(Tamarix chinensis)* and native arrow-weed *(Pluchea sericea)* have replaced cottonwoods as the dominant vegetation type along extensive reaches of the Colorado River (Ohmart et al. 1977). And along the Rio Grande in New Mexico, formerly extensive stands of Fremont cottonwoods are being replaced by saltcedar and Russian-olive *(Elaegnus angustifolia)* (Howe and Knopf 1991).

In Mexico, the Santa Cruz River, which is the major supplier of potable water for the city of Nogales, Sonora, is contaminated by high levels of nitrates and coliform bacteria from the untreated waters of local industries (Castellanos, pers. comm., 1994). Similar problems exist for the San Pedro River, which is contaminated by heavy metals, particularly from copper mining operations near its headwaters in Cananea, Sonora (Moreno 1992). Of the major rivers of northern Mexico, the Colorado River may be the most affected by human insults. The effects of agricultural activities (in particular the use of large quantities of Colorado River water to irrigate vast farmlands such as those of the Imperial Valley, California; San Luis Rio Colorado, Sonora; and Mexicali, Baja California) have combined with urban and industrial contamination to dramatically alter the physical, chemical, and biological characteristics of the Colorado River delta (Sanchez 1991).

For those who have just begun to study riparian ecosystems, background literature on riparian ecology and conservation issues by state is listed in the bibliography. These references will provide a solid foundation for understanding the ecology and value of riparian ecosystems and some of the conservation issues that pertain to them.

Lessons Learned from
Past Riparian Recovery Efforts

Riparian revegetation — which involves planting trees, shrubs, forbs, or grasses to replace species that have been lost — is one of several recovery strategies that have been used to address the decline of riparian ecosystems in the western

United States (Anderson et al. 1978). Other recovery strategies include improving livestock management, installing streambank stabilizing structures, and performing upland treatments. These strategies have been implemented either alone or in combination, with varying degrees of success. In addition, legislation designed to protect riparian ecosystems by maintaining minimum streamflows has been developed during recent years and may play a major role in protecting riparian ecosystems in the years to come (Arizona Rivers Coalition 1991).

Riparian revegetation uses all types of propagules, including cuttings, poles, seedlings, and seeds. When used effectively, revegetation can produce dramatic results, helping to replace lost riparian vegetation and stabilize deteriorating conditions, thereby initiating recovery of the riparian ecosystem. For example, establishing woody vegetation along streambanks can expedite the recovery of damaged riparian areas by slowing or preventing streambank erosion, which provides greater opportunities for other vegetation (e.g., grasses, sedges, forbs, and rushes) to establish (Porter and Silberberger 1961; Miller and Borland 1963; Maddock 1976).

One of the most critical lessons learned from the experiences of past riparian recovery efforts, as well as from recovery efforts that focused on other ecosystems, is the importance of evaluating site conditions to understand current conditions, the extent to which they have declined, and the reasons for their decline (Van Haveren and Jackson 1986; Carothers et al. 1990; Briggs et al. 1994). In general, we are often guilty of jumping to conclusions about the causes of degradation, how we are going to address the causes, and what the end point should be. Only by evaluating site conditions can the information required to develop a realistic and effective recovery plan be collected.

Underscoring this point are the results of a study that evaluated a group of riparian revegetation projects in Arizona (Briggs 1992). This study found that incorporating recovery techniques (in addition to, or in lieu of, revegetation) such as bank stabilization structures, check dams (fig. 1.2), irrigation, and improved land management strategies that address the causes of site decline (either indirectly or directly) was a commonality of the majority of the projects that achieved their objectives. In brief, the ability of these other recovery strategies to overcome the causes of site degradation appears to have a more significant impact on the overall results of the projects than does revegetation. Designing and implementing recovery strategies that address the causes of site decline, of course, requires that the causes of site decline be understood. These results also illustrate why riparian revegetation often produces only marginal results — when revegetation is used alone, the factors responsible for the ini-

Figure 1.2 One of two check dams that were installed in Sheepshead Spring, near Cottonwood, Arizona, to promote alluvial deposition. (Photograph by Liz Rosan, Sonoran Institute)

tial degradation usually hamper or prevent the establishment of artificially planted vegetation. Other important lessons learned from this evaluation of riparian revegetation projects include the following:

> The causes of riparian decline can best be understood by considering the riparian area in the context of its watershed; this requires that reaches upstream and downstream from the degraded riparian area, the tributaries of the drainageway that passes through the degraded riparian area, and the uplands be included in the evaluation process.

> In line with the above, strategies for repairing degraded riparian ecosystems need to take a top-down approach that begins by addressing upland problems. Focusing recovery efforts on the bottomlands and neglecting upland problems may not bring about the intended results. Some of the more effective riparian recovery efforts were focused on solving upland problems, allowing the riparian environment to come back naturally.

Addressing impacts that occur directly in the riparian zone is also an important ingredient for bringing the riparian area back to health.

Riparian ecosystems are resilient. Understanding the potential for natural recovery may, in some cases, eliminate the need for riparian revegetation or other types of streamside recovery efforts. More important, working with natural processes to foster natural regrowth should be the aim of all riparian recovery efforts.

Effective riparian recovery efforts need to be based on a sound understanding of site conditions, with an understanding of water availability, channel dynamics, and soil conditions being the most important.

Generally, riparian revegetation is effective within a fairly narrow range of possible site conditions. On one hand, many degraded areas are too unstable to support the vegetation planted during revegetation; on the other hand, some sites are capable of prolific natural regrowth, making artificial revegetation unnecessary.

An Evaluation Strategy

The dynamic nature of riparian ecosystems, their complexity, and the fact that even riparian ecosystems along the same drainageway can differ substantially from one another in plant composition and hydrologic characteristics make it imperative that recovery strategies be developed on a site-by-site basis. The evaluation should provide the ecosystem manager with a sound understanding of current site conditions, the condition of the site's watershed, how the site's physical and biological characteristics have changed, and the reasons for the changes.

Rather than providing step-by-step instructions, this book discusses general approaches that can be tailored to specific situations. Chapter 2 outlines a strategy that will assist ecosystem managers in understanding the damaged riparian area's watershed, including the importance of understanding the land uses that occur in the watershed and how to gather information that describes the condition of stream reaches upstream and downstream from the riparian site, as well as its tributaries. Chapter 3 evaluates the impacts of human-related disturbances that frequently occur within the riparian zone and describes various strategies for reducing their impacts. Chapter 4 outlines some important considerations for determining the potential that a damaged riparian area will experience strong natural recovery, possibly making riparian revegetation or other types of streamside manipulations unnecessary. Chapters

5, 6, and 7 present criteria for evaluating water availability, channel stability, and soil salinity, respectively. Finally, Chapter 8 provides some important considerations for organizing the acquired information into a viable riparian recovery plan. The epilogue clarifies the premises and the goals of this book.

Defining Some Important Terms

Although a glossary is provided at the end of the book, definitions of several terms are highlighted here to clarify their meaning from the outset.

Ecosystem Manager

For this guidebook, the term *ecosystem manager* has a very broad, inclusionary meaning, referring to anyone who is involved with evaluating an ecosystem with the intent of developing a plan for improving its condition.

Bottomland Ecosystem versus Riparian Ecosystem

Most biologists and ecologists use the term *riparian ecosystem* to describe "vegetation, habitat, or an ecosystem that is associated with bodies of water (streams or lakes) or is dependent on the existence of perennial, intermittent, or ephemeral surface or subsurface water drainage" (Arizona Riparian Council 1996). In this respect, the term *riparian* has a fairly broad meaning, and the extent of a riparian area is determined by several parameters, including water availability, topography, and vegetation characteristics. However, for many hydrologists the term *riparian* has a more narrow meaning, referring only to the area immediately adjacent to a natural watercourse.

The term *bottomland* avoids this disparity because it is used only to describe a specific type of landscape: bottomland is defined as low-lying, nearly flat land along a watercourse. Bottomlands are formed of terraces, floodplains, and all surfaces lower than the floodplain, including the channel. Bottomland ecosystems are the biotic communities associated with the bottomlands of rivers, streams, lakes, and other landscape settings where the availability of water is greater than in the surrounding uplands.

In many situations, use of the term *bottomland ecosystem* is preferable because it avoids the confusion that the term *riparian ecosystem* can cause. Nevertheless, *riparian ecosystem* is used more widely, and because this book draws heavily from the biological perspective, *riparian ecosystem* will be used here.

Healthy versus Degraded Ecological Condition

In conservation literature, the term *ecological condition* (or *ecological health*) is often used, yet rarely defined. It is a nebulous term whose definition often depends on management objectives. For example, an ecologist developing a management plan to preserve spotted owl *(Strix occidentalis)* habitat in the Pacific Northwest will have an entirely different set of criteria for determining the ecological condition of a forest ecosystem than would a forester cruising timber.

The term *healthy* as used here describes a riparian ecosystem with a high diversity of native plant species, and one that has not experienced dramatic changes in structure (e.g., age- or size-class, composition, diversity, biomass) or functional elements (e.g., changes in erosion rates, flow regime, groundwater, channel morphology, productivity) in comparison to its pre-Anglo-settlement condition. A degraded or damaged ecosystem is defined by the reverse of the above.

The word *degrade* is also used differently by the various science disciplines. Ecologists often use *degrade* to describe a reduction in the ecological condition of an ecosystem; hydrologists to describe adjustments in channel profile. Here, *degrade* is used mainly to describe a decline in ecological condition. However, in chapter 6, *degrade* is also used to describe channel conditions: the emphasis in this chapter on hydrology and drainageway conditions made the overlap of definitions unavoidable.

Recovery

Generally, there are three possible end points for efforts undertaken to improve the condition of an ecosystem that is in less than pristine condition: restoration, rehabilitation, and replacement/reallocation (Bradshaw 1988; Aronson et al. 1993). Restoration is an attempt to create an ecosystem exactly like the one that was present prior to disturbance. Rehabilitation creates an ecosystem that is similar (but not identical) to the ecosystem that was present prior to the disturbance. Replacement or reallocation strategies generally do not attempt to restore an ecosystem to its predisturbance condition but instead are intended to replace the original ecosystem with a different one. The term *recovery* is used when the objective is unclear: *recovery* may be used to refer to any type of project (restoration, rehabilitation, or replacement) that is undertaken to improve the condition of degraded riparian ecosystems.

2/ Considering the Damaged Riparian Area from a Watershed Perspective

An important lesson learned from past riparian recovery efforts is the need to evaluate the condition of the damaged riparian area from a watershed perspective that takes into consideration the uplands, upstream and downstream reaches, and tributaries. This is crucial because the structure and processes of lotic ecosystems, more than any other type of ecosystem, are determined by their connection with adjacent ecosystems. Since riparian ecosystems are in the bottomlands of a watershed, changes in how sediment and water run off of surrounding landscapes impact them most. A disturbance in any part of a watershed will create disequilibrium that will ripple through many ecosystems within the watershed, possibly influencing the condition of the watershed's riparian ecosystems for many years.

Ecosystem managers must therefore avoid the myopic approach of developing recovery strategies that are based solely on an evaluation of the immediate degraded riparian site. Evaluating only isolated components of a watershed (e.g., a specific stream reach) will be ecologically incomplete, often failing to provide the information needed to fully understand why the condition of the riparian ecosystem has declined.

The evaluation process should also be broadened from a temporal perspective. Collecting information describing past conditions will provide a picture of how the riparian area has changed and some of the potential reasons for the changes.

This chapter provides a checklist that covers some of the basic information that ecosystem managers should understand from the start about the damaged riparian area's watershed and the drainageway that passes through it. Although general in nature, the information gathered here will form the foundation for more detailed evaluation work that may be required when the evaluation process moves to focusing specifically on the degraded riparian area. This type of evaluation approach — going from a broad perspective to specifics, as well as from gathering general information to more detailed information — is being used to evaluate the condition of lower Rincon Creek, near Tucson, Arizona (case study 1).

Case Study 1
Rincon Creek
Determining the Causes of Degradation

Rincon Creek drains a significant portion of the west-facing slopes of the Rincon Mountains near Tucson, Arizona. Only 40 years ago, the lower reach of this drainageway experienced aboveground flow most of the year and was dominated by a large Arizona walnut *(Juglans major)* and mesquite *(Prosopis velutina)* woodland. Today, this reach experiences flow only in direct response to significant rainfall and is dominated by such species as burro-weed *(Aplopappus* sp.), burrobrush *(Hymenoclea monogyra),* ragweed *(Ambrosia* spp.), and jimsonweed *(Datura stramonium)* (fig. 2.1).

This reach of Rincon Creek is being evaluated to find out why its condition has changed so dramatically and what can be done to bring it back to health. The bottomlands of lower Rincon Creek are currently grazed by cattle, and preliminary evaluations of this area seemed to point to overgrazing as the major cause of decline. However, when the evaluation was broadened to include other considerations, the more pivotal role of other impacts became obvious.

Interviews with long-time residents, county records, and old photographs reveal an extensive history of agriculture that goes back to the early 1930s. To make way for the cultivation of alfalfa and barley, the streamside vegetation communities were removed; pumping records show that these crops relied heavily on irrigation for their survival. Current records of water wells reveal groundwater depths that approach 25 m, far below what is suitable for most phreatophytes. Including Rincon Creek tributaries in the evaluation process and comparing aerial photographs from the mid 1930s to more recent times revealed losses in streamside vegetation communities and the construction of retention basins along many of the tributaries of Rincon Creek, which prevent the tributaries from directly contributing to flow in Rincon Creek.

Data collected from crest stage gauges and scour and fill chains installed along a significant portion of lower Rincon Creek portray an unstable environment that experiences significant aggradation and flashy flow events. In addition, a significant portion of the watershed of Rincon Creek is likely to be extensively developed in the near future, threatening to further heighten the stream channel's instability.

Figure 2.1 Lower Rincon Creek near Tucson, Arizona, as it appears today. (Photograph by Liz Rosan, Sonoran Institute)

Rincon Creek provides a nice example of how a damaged riparian site is being evaluated so that an effective reparation strategy can be developed. The Rincon Creek example is referred to in various parts of this book.

Taking Advantage of Available Information

Taking advantage of documented information about the degraded riparian area can save ecosystem managers a lot of time, energy, and money. At least some background information is likely to be available. Federal and state agencies (see bibliography, list of archives), university libraries, and local private conservation organizations are good places to begin the search. Many areas have been directly studied as part of management plan development, scholarly work, resource inventories, environmental impact assessments, and so forth.

If an area has not been directly studied, at least indirect data are likely to be available. Maps, field data, archival sources, and surveys of bridge and pipeline crossings can provide information describing changes in channel form and independent geomorphic watershed variables. Conservation plans, historical records, and fishing and hunting records can provide information describing past vegetation, wildlife, and fish characteristics.

Aerial photographs can aid tremendously in understanding the past condition of the watershed and its associated riparian ecosystems. Comparing historical photographs to current versions is a relatively quick and inexpensive way to understand the extent to which the watershed has changed since the mid 1930s (when aerial photographs became available for much of the United States) and the reasons for the changes. Such a comparison can provide information on rates of agricultural expansion, urbanization (housing, roads, etc.), timber harvesting, grazing activities, and a plethora of other land use activities, which in turn can help explain the decline of riparian habitat. Numerous texts are available on aerial photograph interpretation: two examples that focus on natural resource issues are Verstappen 1983 and Way 1978.

Universities and most federal and state agencies have libraries of aerial photographs. The U.S. Geological Survey maintains a central repository of photographs from the United States and many other parts of the world in Sioux Falls, South Dakota. Aerial photographs taken prior to 1945 can be found at the National Archives in Washington, D.C. In addition, most large urban centers have private firms that produce and sell their own aerial photographs.

Getting to Know the Watershed

A watershed is the area that collects and discharges runoff for a point on a stream. The watershed is therefore defined by the stream that drains it. The terms *catchment* and *drainage basin* are often used synonymously with *water-*

shed, whereas the term *drainageway* refers to principal areas of water accumulation (i.e., channels).

Climate

The climate of the watershed is one of the primary factors (with geology being the other) that affect the pattern of streamflow, shape of the landscape, and density and diversity of vegetation communities. Understanding the climate of the degraded riparian area's watershed and how it may have changed is an important step toward understanding the reasons for the vegetation and hydrologic changes that have occurred. Climate is often classified based on the moisture budget and potential evapotranspiration, requiring the measurement of temperature, relative humidity, and precipitation.

Collecting precipitation data is important because relations between precipitation and runoff can be developed, using as parameters some of the factors that affect runoff. Such a relationship can be useful because it can be extrapolated to areas that lack hydrologic data or be used to estimate runoff for the area of interest. However, the relation is often unsatisfactory, unless other parameters (such as storm frequency, initial soil-moisture condition, storm duration, and time of year) are also taken into consideration.

Precipitation data are often available from various agencies and institutions. Global precipitation data are available at the National Climatic Data Center in Asheville, North Carolina; regional offices have data more specific to a certain part of the United States. In addition, each state has an Office of the State Climatologist that collects and distributes precipitation data for its associated state. Precipitation data for large urban areas are often available in the "World Weather Almanac," a publication that pulls together precipitation data from around the world. In addition, precipitation data are often collected for areas managed by the National Park Service, Forest Service, and Bureau of Land Management.

Watershed Size

The size of the watershed area is an important descriptor because it influences water yield and the number and size of streams. Watershed boundaries can be located using the contour lines on a topographic map. After the watershed is delineated on a map, its size can be determined in a number of ways. A planimeter can be used to measure the area directly. Watershed size can also be measured by placing a square or dot grid over the area that is delineated on the map; the number of squares (or fractions of squares) or dots that lie in the watershed area are then counted. Methods for measuring watershed size are reviewed in detail in documents such as Gordon et al. 1992.

Land Uses in the Watershed

Understanding the watershed's land use history is a significant step toward understanding the trend of the landscape, the reasons behind the decline of its riparian ecosystems, and the potential for recovery. Going back to the time when the watershed was first influenced significantly by Anglo-American settlement is often sufficient for understanding how the watershed has been most affected by human activities. For the Rincon Creek watershed, for example, information describing its land use history is being collected from the turn of the century, when Anglo-American settlement patterns and livestock use became significant.

Ecosystem managers should also consider how land use activities in the watershed may change in the future. Although currently in a rural setting, the Rincon Creek reach that is being evaluated for recovery (case study 1) passes through landscapes that are likely to experience extensive development in the near future. To be effective, the Rincon Creek recovery project needs to take into consideration how such a dramatic change in land use will affect factors such as sediment and water runoff, groundwater levels, and channel stability—factors that can significantly affect the outcome of a riparian recovery effort. In addition, ecosystem managers may have to design the recovery project so that it meets the aspirations of the developing community. This, of course, affects not only the riparian recovery strategy that is ultimately developed, but the overall objective of the project as well.

Drainage Density

Drainage density is an expression of how a watershed is dissected by its drainages (Toy and Hadley 1987) and is calculated by dividing the total stream length for the watershed by the catchment area:

$$D_d = \frac{\Sigma L}{A}$$

where D_d is the drainage density, ΣL is the total channel length, and A is the catchment area. Channel length is commonly measured directly off of

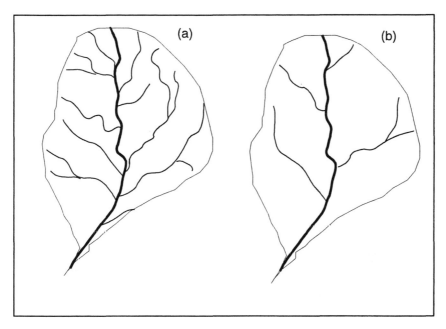

Figure 2.2 Watersheds with (a) high and (b) low drainage densities. (Illustration by Mark Briggs)

topographic maps. However, using up-to-date aerial photographs is more accurate.

Watersheds with low drainage densities are characterized by a broadly divided network of drainages with long channel lengths and shallow slopes (fig. 2.2b). In comparison, watersheds with high drainage densities have a more finely divided drainage network with shorter lengths and steep slopes (fig. 2.2a). The greater the drainage density, the shorter the length of hillslopes and the shorter the length of overland flow. Overland flow is the distance that water flows over the ground surface before it enters a stream channel (Horton 1945).

A watershed's drainage density can reveal useful information about the watershed and the processes that are shaping it. Drainage density is influenced by climate, soil, geology, and vegetation characteristics. Watersheds located in semiarid areas generally have higher drainage densities than those located in more mesic climates. A comparison of drainage densities over time can provide clues describing the general trend of the watershed. For example, an increase in drainage density may be an indication that the watershed is reacting to a disturbance (e.g., timber harvesting, overgrazing by livestock) that has changed the way that sediment and water run off of the watershed surfaces.

Riparian Vegetation Characteristics

As part of the initial evaluation process, ecosystem managers should also consider the condition of the riparian ecosystems upstream and downstream from the degraded riparian site. Such information can be invaluable for determining the extent and causes of decline. If the riparian communities upstream and downstream from the degraded riparian site are in relatively good ecological health, then the causes of decline are probably due to local rather than regional impacts. By contrast, deterioration of riparian communities along a significant portion of a drainageway can be an indication of impacts that are affecting large areas (e.g., groundwater decline, upland disturbances). Looking at the riparian plant communities along a significant portion of the drainageway will also give ecosystem managers the opportunity to note the locations of desirable seed sources, an important piece of information for predicting the potential that the damaged riparian site will recover on its own (see chapter 4).

Getting to Know the Stream

The germination and growth of riparian vegetation communities are intricately related to river discharge, movement of the channel, and development of the floodplain (Everitt 1968; Hendrickson and Minckley 1985). To be effective, riparian recovery strategies have to be based on a sound understanding of the stream that passes through the degraded riparian area. This section introduces some basic channel and streamflow descriptors that should be understood by the ecosystem managers as part of the initial evaluation effort. Chapter 6 discusses in detail methods for evaluating channel geometry, streamflow, bank erosion rates, and sediment movement.

Is the Channel Bedrock or Alluvium Controlled?

Stream channel form is generally categorized as being either alluvium or bedrock. Bedrock channels are predominantly controlled by geology; alluvial channels by streamflow. In bedrock channels, streamflow is confined by rock outcrops, and changes in channel morphology generally occur slowly over a long time period. In comparison, alluvial channels are characterized by channel beds and banks composed of materials transported by the river under the current flow conditions. Alluvial channels are therefore free to adjust their gradient, dimensions, and shape.

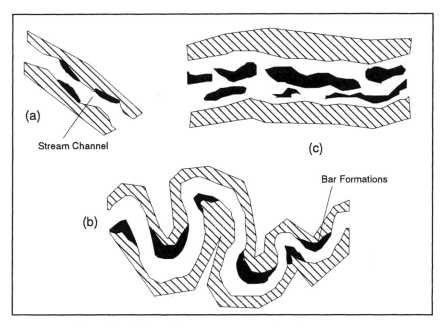

Figure 2.3 Examples of channel patterns: (a) straight, (b) meandering, (c) braided. (Illustration by Mark Briggs)

Recovery options that will be effective for alluvial channels may not be for bedrock channels, and vice versa. For example, constructing check dams may not be a wise option in an alluvial channel because of the potential for explosive lateral erosion around the sides of the dam—something that is not as likely when the check dam is anchored in bedrock. Furthermore, the types of evaluation required to determine appropriate recovery will be different for the two types of channels. For example, evaluating channel stability is not necessary for bedrock channels, which are inherently more stable than alluvial channels.

Channel Pattern and Shape

Channel pattern refers to the top or plan view of the stream and its alignment, and channel shape relates to the cross-sectional view of the channel. Channel patterns are commonly described as being braided, meandering, or straight (fig. 2.3), although intermediate channel patterns exist that can make distinguishing between straight, meandering, and braided channels difficult.

Leopold and Wolman (1957, p. 40) noted, for example, that "there is a gradual merging of one pattern into another." Nevertheless, looking at both of these parameters and how they have changed over the years can provide ecosystem managers with valuable information regarding the stream channel's stability and the appropriateness of using revegetation or other types of recovery efforts (see chapter 6).

Channel Gradient

Channel gradient is influenced by numerous factors, including degradation and aggradation rates, changes in streambed material, and local topographic changes. The slope of a stream can be measured along a short reach or averaged over a much longer segment. A simple method is to determine the mean channel slope (S_c), which is given by

$$S_c = \frac{\Delta_e}{L}$$

where Δ_e is the elevation at a stream's source minus its elevation at the mouth, and L is the length of stream. These measurements can be taken off of a topographic map. Surveying, of course, would provide greater accuracy.

Longitudinal Profile

The longitudinal profile of a stream describes how a stream's elevation changes as it moves from its upper reaches to its lower reaches. The longitudinal profile of a stream is often displayed on a graph with the y-axis being elevation and the x-axis being distance from mouth (fig. 2.4). Describing the longitudinal profile of the stream that passes through the degraded riparian area provides meaningful information for understanding the characteristics of channel reaches both upstream and downstream from the degraded riparian site.

Many streams exhibit a longitudinal profile that has a general concave shape, with the slope decreasing as the stream moves from its upper to lower reaches. Schumm (1977) noted that a stream that shows a concave longitudinal profile is often associated with an increase in discharge and a decrease in sediment size as it moves to the lower elevations. A concave profile often does not develop for drainages that obtain the majority of their flow from the headwater regions and do not increase in discharge downstream (Petts and Foster 1985). A stream's longitudinal profile is also affected by local topography, changes in bed materials, and knickpoints.

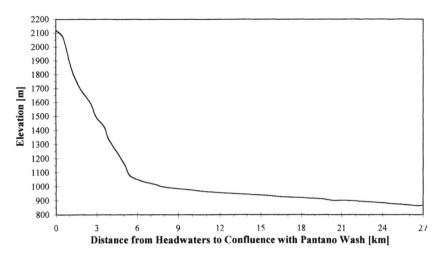

Figure 2.4 The longitudinal profile of Rincon Creek near Tucson, Arizona — typical of many streams in the southwestern United States in having a relatively steep gradient at its headwaters and an increasingly gentle gradient as the drainageway approaches the desert floor. (Illustration by Mary Schmid, Rincon Institute)

Streamflow Parameters

Streamflow characteristics, such as annual discharge and the frequency and timing of flood events, play a significant role in determining the vegetation characteristics of riparian communities. Understanding these and other streamflow parameters and how they may be changing also will help in understanding why the riparian area is damaged and what can be done to repair it. For example, an abrupt change in streamflow (e.g., a channel reach changing from perennial to ephemeral flow over a ten-year period) may be an indication that the area is being influenced by land use impacts that will have to be taken into account in the design of the riparian recovery project. Knowledge of the magnitude and recurrence frequency of floods is needed for the design of check dams, streambank protection devices, and other in-stream structures. Combining information describing streamflow with channel morphology data can also help ecosystem managers compare potential revegetation sites in their vulnerability to flood scour.

Such basic streamflow descriptors as the percentage of time the stream experiences aboveground flow are easily determined for streams that have been studied (e.g., gaged streams); however, such information may be lacking for streams in remote areas. Gathering flow and sediment data for ungaged streams is discussed in chapter 6.

The terms *ephemeral, intermittent,* and *perennial* are used to give an indication of the approximate time that a given stream is expected to experience aboveground flow. An ephemeral stream is a stream or reach of a channel that flows only in direct response to precipitation in the immediate locality, and whose channel is at all times above the zone of saturation. This means that an ephemeral stream will lose water to the streambed, ensuring a decrease in flood discharge downstream in the absence of significant tributary inflows. An intermittent stream is a stream or reach of a channel that flows only during certain times of the year (e.g., when it receives water from springs or seeps). A perennial stream is a stream or reach of a channel that flows continuously, or nearly so, throughout the year and whose upper surface is generally lower than the top of the zone of saturation in the areas adjacent to the stream.

Streamflow parameters such as total volume, degree of variability, seasonal distribution, and average, maximum, and minimum discharge are important basic descriptors that can help ecosystem managers understand the current hydrologic condition of the area and how it has changed. Statistical measures of the annual, seasonal, and daily streamflow patterns can reveal differences between drainageways or between different reaches of the same stream, as well as changes due to natural trends or human activities (Gordon et al. 1992).

Mean annual runoff represents the difference between annual evaporation and precipitation. This parameter is important because it provides an indication of the average amount of water that the stream's watershed provides on an annual basis. Mean annual flow is commonly reported as either total volume or average discharge. For streamflow records tabulated on a daily basis, mean annual flow (as an average discharge) can be computed by averaging the daily data for the water year. Mean annual flow can also be estimated by dividing the average annual volume (in millions of cubic meters) by 31.5576.

Monthly streamflow statistics will reveal seasonal variations in discharge, which are controlled by climate and channel and watershed characteristics. Seasonal variation in streamflow can be used to classify the stream's flow regime. For riparian revegetation projects, monthly statistics can be used to indicate potential water availability or shortage at various times of the year. For example, ecosystem managers may want to avoid revegetating a damaged riparian area during the time of year when streamflow is at a minimum or, in the case of ephemeral and intermittent channels, nonexistent.

Daily statistics provide a finer resolution for computing seasonal streamflow patterns over the year. Daily streamflow data can be used to develop an average annual hydrograph. Points on the hydrograph represent the average value for each day, calculated as an average over all the years for the period of interest, or simply for the time period for which data are available.

Is Streamflow Regulated or Natural?

By altering natural patterns of flooding, sedimentation, and groundwater fluctuations, impoundment structures (e.g., dams, retention basins) can affect the condition of riparian ecosystems (Buma and Days 1975), both upstream and downstream from the impoundment structure (Wolman 1967; Hagen and Roberts 1972; Gill 1973; Kerr 1973). The disappearance of riparian habitat along the lower Colorado River (Ohmart et al. 1977), Salt River (Fenner et al. 1985), and lower Verde River (McNatt ct al. 1980) in Arizona and the Rio Grande in New Mexico (Howe and Knopf 1991) has been attributed, at least in part, to the reduction of overbank flows caused by the buffering effect of impoundments.

The effects of impoundment on natural streamflow can greatly reduce the ability of many riparian species to successfully propagate. High spring flows, after being captured in reservoirs, are released over a much longer time period. Buffered spring flows may not have sufficient power to deposit reworked alluvium and seed in areas sufficiently removed from the base flow area. As a result, seeds often germinate in areas vulnerable to scouring and may not survive subsequent flooding. This situation may produce riparian vegetation patterns characterized by a predominance of old trees with little seedling development. In fact, in some locations (e.g., the lower Colorado River), impoundment structures have altered natural flows so dramatically that revegetation may be one of the few options left for reintroducing native riparian species (Briggs 1992).

Ecosystem managers should determine whether impoundment structures have been constructed across any portion of the stream that passes through the degraded riparian site. Maps and aerial photographs can be used to locate impoundment structures. Ground surveys are, of course, more accurate, providing information that may be absent from maps and aerial photographs, especially if the maps and aerial photographs are not current.

$3/$ Impacts within the Riparian Zone

Evaluating the effects of land use activities that occur within the immediate riparian environment and then reducing or eliminating them if they are causing damage should be one of the main priorities of the recovery effort. This strategy is often called "passive restoration" because it does not involve the use of revegetation, in-stream structures, or any other type of active manipulations (Kauffman et al. 1995). Once the causes of decline are addressed, the ecological resiliency of riparian ecosystems takes over, often leading to the dramatic recovery of the system. This approach, however, has often been ignored in favor of active recovery strategies that address only the symptoms of site decline (e.g., using in-channel structures to control bank erosion).

Agricultural activities, flood control measures, livestock grazing, and recreational activities are just a few examples of impacts that often occur within the riparian zone. The impacts of livestock grazing and recreational activities and suggestions for how they can be reduced will be considered in this chapter. Competition from nonnative vegetation species and the impacts of wildlife are also discussed, particularly from the perspective of how these factors can influence the results of revegetation.

Livestock

Livestock Grazing and Riparian Ecosystems

Livestock can produce long-lasting detrimental impacts on riparian ecosystems (Behnke and Raleigh 1978; Meehan and Platts 1978; Busby 1979; Duff 1979; Elmore and Beschta 1987). Many land managers point out, however, that it is not livestock grazing that damages riparian ecosystems, but livestock overgrazing. Overgrazing is likely to occur when land is grazed repeatedly without intervening periods of rest that are of sufficient length for recovery. Overgrazed conditions can also result when sensitive land (e.g., land characterized by steep slopes, shallow soils, low vegetation densities) is grazed severely just once.

The effects of overgrazing on riparian ecosystems are well documented. High grazing intensities (high numbers of livestock over a long period of time) can alter plant composition (Elmore 1989; Platts et al. 1989; Svejcar 1989; Archer and Smeins 1991), affect soil infiltration rates (Gifford and Hawkins 1978), and increase soil erosion (Swanson et al. 1987; Platts 1989). Overgrazing can also negatively influence several parameters associated with stream habitat, including the reduction or loss of streambank undercuts (Platts 1989) and a decline in water quality due to increased turbidity (Elmore 1989) and fecal contamination (Platts 1989).

Generally, riparian ecosystems offer more forage of higher quality (i.e., a higher proportion of green to dead plant material and a higher proportion of leaves to stems) and greater amounts of shade and water than do upland ecosystems. Such characteristics tend to attract livestock, particularly during the hot, dry months in arid and semiarid climates. As a result, riparian areas often receive a disproportionate amount of use in comparison to their size. For example, in eastern Oregon, 50% of the cattle that were studied were located on 5% of the land area; forage utilization was 75% on riparian meadows and only 10% on uplands (Gillen et al. 1985). And although streamside vegetation makes up only about 10% of the total vegetation found in meadows in northern California, it provides about 90% of the usable forage for livestock (Svejcar 1989).

The results of several investigations on the simulated effects of cattle grazing on herbaceous plants at low-elevation and high-elevation sites indicate that reduction of plant growth by soil compaction was a relatively consistent effect among the sites that were studied (Clary 1995). An investigation of the trampling, defoliation, and excreta effects of sheep on a perennial ryegrass–white clover (Lolium perenne–Trifolium repens) pasture found that the benefits of increased nitrogen from sheep excreta were outweighed by the negative effects of trampling and defoliation that occurred when stocking rates were increased (Curll and Wilkins 1982, 1983).

The effects of herbivory on riparian ecosystems vary from one riparian site to the next (Green and Kauffman 1995). Certainly, the number of livestock, length of grazing period, and length of time the riparian area is allowed to rest between grazing periods have a significant effect on how long a riparian area can sustain grazing without deteriorating. Several abiotic factors, including soil, stream channel, hydrologic, topographic, and climatic characteristics, also play important roles in determining the susceptibility of riparian areas to damage from livestock grazing. Generally, a riparian ecosystem is more susceptible to livestock damage when it is surrounded by land that is steep and rocky, when surrounding nonriparian forage is less palatable than the riparian forage, when the herd is composed of cows with calves as opposed to yearlings, when animal distribution is not maintained by herding, or when the grazing season is long (Swanson 1988).

For additional background information on the management of livestock in riparian areas, consult Kauffman and Krueger 1984, Skovlin 1984, and Chaney et al. 1990.

Evaluating the Impacts of Livestock

Simply because livestock are in a damaged riparian area does not always mean they are the principal cause of site decline. The extent to which livestock are contributing to site decline needs to be determined so that recovery strategies do not address only minor impacts while leaving the major causes of site decline unsolved. However, making such an assessment is not always straightforward. The decline in the riparian area can be caused by the combined effects of numerous land use activities on any part of the watershed or by a perturbation that may be affecting the condition of a site years after the activity has ceased.

The extent to which livestock are contributing to site decline can be better understood when the following parameters are defined: (1) the current season of use and grazing intensities, and how these have changed with time; (2) the current condition of the damaged riparian site, as well as the riparian communities upstream and downstream from the site, as described by vegetation, hydrology, and stream channel characteristics, and how these characteristics have changed over time; and (3) the watershed's land use history. At Rincon Creek (case study 1; see chapter 2), for example, historical photographs, aerial photographs from the mid 1930s, and interviews with long-time residents revealed that agricultural activities, not livestock use, had contributed most significantly to the decline of the riparian ecosystems in this area despite continued livestock grazing and the cessation of agricultural activities.

Improving Livestock Grazing
Management Strategies

Livestock management should be changed if livestock are contributing to riparian decline. Working with ranchers to develop sustainable livestock management plans has produced startling results. For example, several overgrazed riparian areas in the Tonto National Forest have dramatically recovered as a result of Forest Service personnel working with grazing permittees to reduce grazing intensities and establish designated areas for grazing (Alford 1993). Biologists working for the U.S. Bureau of Land Management (BLM) consider livestock grazing the most significant manageable impact on riparian ecosystems in Arizona, and in recent years, BLM has worked with ranchers to develop specialized grazing plans to improve the condition of overgrazed riparian areas (Hall and Bammann 1987).

Altering livestock management is preferable to revegetation because it addresses the causes of site degradation. The primary objective of developing a specialized livestock grazing management strategy is to produce a situation where livestock are using the land in a sustainable manner. There are, however, many ways by which an unsound grazing management strategy (one that is leading to ecological deterioration) can be changed. The type (cattle or sheep), class (calves, steers, or cows), and number of livestock, as well as the intensity of use, can be changed. Herding can be introduced, fences installed, or upland water sources developed. All, none, or a combination of these strategies and others may produce positive results in any given situation.

As with all recovery strategies, developing improved grazing management requires a clearly defined objective that describes the desired condition of the riparian area (Chaney et al. 1990). Developing such an objective requires that ecosystem managers understand the complexity of the riparian ecosystem (Green and Kauffman 1995): the unique characteristics of the site, the watershed, the drainageway, the terrain, the class or kind of livestock, and the management capabilities and objectives of the permittee or livestock operator (Chaney et al. 1990).

The management strategy that is ultimately developed needs to be tailored to the unique characteristics of the riparian site. At the very least it should provide sufficient rest to encourage plant vigor, regeneration, and energy storage, and it should ensure that vegetation will regenerate in sufficient strength to protect streambanks during times of high flow. In addition, the management strategy should also ensure that grazing does not occur when streambanks are most vulnerable to trampling. Chaney and coworkers (1993) noted that steep, wet streambanks with very little vegetation cover are highly vulnerable to the effects of trampling.

Moir (1989) and Pendleton (1989) discuss the criteria used by the U.S. Forest Service and the U.S. Soil Conservation Service, respectively, to evaluate range condition, and Renner and Allred (1962) review strategies for classifying rangeland for conservation planning.

The following livestock management strategies have been used either alone or in concert with one another with varying levels of success.

Focusing on the Uplands prior to the Bottomlands. As with other recovery strategies, improving livestock management on the uplands prior to focusing on the bottomlands is critical (Platts and Raleigh 1984; Elmore 1989). Dramatic riparian improvement can be accomplished through efforts that focus solely on improving deteriorated upland conditions (Briggs 1992). At Sheepshead Spring (case study 2), for example, most of the effort to improve the bottomland environment actually focused on improving the condition of upland plant communities.

Develop Riparian Pastures as Separate Managed Resources. One of the reasons why livestock (particularly cattle) have often contributed to riparian decline is that the grazing strategies were not developed to sustain riparian ecosystems but instead primarily to increase the production and vigor of upland grasses (Platts and Nelson 1985b). Riparian and nonriparian ecosystems were combined into one management unit, which often resulted in the disproportionate use of riparian ecosystems (Heady 1975; Platts and Wagstaff 1984; Platts and Nelson 1985b). A grazing strategy developed for upland grasses, for example, may result in severe overgrazing of riparian grasses, forbs, and shrubs.

One solution to this problem is to develop a special riparian pasture that includes the riparian corridor along with a portion of the uplands. The special pasture would be located within an allotment but would be managed separately (Platts and Nelson 1985b). This strategy has produced some positive results. Platts and Nelson (1985b) noted that by adapting the sizes and shapes of pastures to fit the ratio of riparian forage use versus upland forage use and site characteristics, it may be possible to graze the riparian environment without damaging it.

Rest-Rotation Strategies. Rest-rotation strategies involve removing an area from grazing use while the remainder of the pasture continues to support grazing. This strategy requires moving livestock from one area to another.

Rest-rotation strategies produced impressive results when they were implemented in several grazing allotments in the Tonto National Forest in central Arizona. For these allotments, management was changed from continuous, season-long grazing to a five-pasture, rest-rotation system providing high-intensity, short-duration grazing and spring-summer rest two out of every three years. In one study area, more than 400 cottonwoods and 1,280 willows per hectare, measuring 6.4 mm to 657 mm in diameter, had recovered naturally in an area where previously there were none (Chaney et al. 1990). Similar recovery results were experienced at Date Creek northwest of Wickenburg, Arizona, when year-round grazing was changed to a management strategy in which 400 head of cattle are grazed from November to March along a two-mile stream reach, which is then rested the remainder of the year (Dagget 1994).

As with the other strategies presented in this section, rest-rotation strategies have certain limitations and should not be used in all situations. For example, Meehan and Platts (1978) noted that rest-rotation strategies may cause trampling damage that can lead to streambank damage and increased erosion rates.

Fencing. Improving livestock grazing management to alleviate pressure on the bottomlands often requires some fencing of stream areas (Anseth 1983; Manci 1989). Of the range management strategies that are in common use today, fencing the riparian area to limit livestock use (or to exclude livestock altogether) offers the most protection and the best chance for recovery in the shortest period of time (Duff 1979; Keller et al. 1979; Platts and Wagstaff 1984). Fences can also be used to divide the riparian pasture into smaller management units so that at least a portion of it can be rested every year.

There are some problems with installing fences, and in general, fences should not be seen as a panacea for solving problems associated with managing livestock in riparian areas. The cost of fencing can be prohibitive. For example, Platts and Wagstaff (1984) estimated that it would cost BLM more than 90 million dollars to install fencing along stream fisheries within their jurisdiction, and another 9.4 million dollars for maintenance over a twenty-year period.

In addition, ecosystem managers should evaluate the potential side effects that fencing the riparian corridor can have on wildlife (especially large mammals like elk, moose, and deer) that may need to cross the fence line. Using gates, decreasing fence height, or even laying down the fence during certain times of the year are some practical strategies that can lessen the impacts of fences on wildlife.

Upland Water Development. Developing water sources away from the riparian ecosystem is another strategy that can reduce the time that livestock spend in riparian areas. However, the ability of upland water development to influence livestock behavior is debatable. Stoddart et al. (1975) found that well-

Case Study 2
Sheepshead Spring
Focusing on the Uplands before the Bottomlands

Sheepshead Spring is a small, perennial stream in the Coconino National Forest near Cornville, Arizona. Prior to significant human activities that negatively affected the condition of upland and bottomland ecosystems, the Sheepshead Spring riparian environment was dominated by Goodding willow *(Salix gooddingii)* and Fremont cottonwood *(Populus fremontii)*. By 1981, however, the ecological condition of this riparian area had declined significantly because of years of livestock overgrazing. Much of the channel bed had eroded to bedrock, and streamside vegetation consisted predominately of annual grasses, arrow-weed *(Pluchea sericea),* and spike-rush (*Eleocharis* spp.). However, scattered old-growth cottonwood and willow persevered 2 km upstream from the site.

The objective of the recovery effort was to improve the ecological condition of the riparian ecosystem. Revegetation was not performed. Instead, livestock were excluded from the bottomlands, and improved livestock management schemes on the uplands were implemented. In addition, two check dams were constructed across the width of the channel to promote alluvial deposition (fig. 1.2).

Vegetation regeneration eight years after the completion of work was considerable. Numerous cottonwood and Goodding willow seedlings had reestablished (growing to more than 2.5 m tall), and a dense, well-developed understory had developed that included such species as cattail (*Typha* spp.), bristlegrass (*Setaria* spp.), and Kentucky bluegrass *(Poa pratensis)*. Today, an extensive riparian forest has reestablished just downstream from the check dams—another testimony of the resiliency of riparian ecosystems (fig. 3.1).

According to the project manager, the key to success was working closely with the permittee to develop grazing patterns that would improve the ecological condition of the uplands (Goodwin, pers. comm., 1990). The results experienced at Sheepshead Spring underscore the importance of identifying and then addressing the causes of site decline, allowing the natural healing processes of the riparian ecosystem to take over.

Figure 3.1 Dramatic natural regeneration of riparian species in Sheepshead Spring. (Photograph by Liz Rosan, Sonoran Institute)

spaced watering points can benefit animal health and reduce range deterioration. In contrast, Bryant (1982) observed that alternative water sources away from the riparian zone did not significantly affect livestock distribution. Brown and Karsky (1989) provide additional information on developing watering facilities for livestock.

Working with the Permittee. Working closely with permittees (livestock owners with grazing permits) can greatly improve the effectiveness with which new management plans are implemented. Alford (1993) noted that cooperation between the grazing permittee, the Forest Service, and range conservationists was a crucial ingredient in developing and implementing improved livestock management in the Tonto National Forest, central Arizona. Crucial aspects of an improved livestock management plan such as opening and closing gates, development of off-site water and forage facilities, and fence and stopgap maintenance may be impossible without the cooperation of the permittee (Hall and Zarlingo, pers. comm., 1990).

A close relationship with the permittee will also be beneficial if a riparian revegetation strategy is used. Permittees can help ecosystem managers inspect plantings, monitor irrigation operation, and assist with postplanting procedures, all of which will greatly improve the effectiveness of revegetation.

Keeping Management Plans Flexible. Management plans should be flexible and capable of adjustment to annual precipitation fluctuations, the rate of ecosystem recovery, and the degree of degradation. In some instances, grazing intensity can be increased after riparian vegetation shows signs of recovery (Hall, pers. comm., 1990). Elmore and Beschta (1987), however, cautioned against changing plans too quickly to take advantage of unused forage. They observed that "unused forage" is necessary to maintain the integrity of the system. Increasing livestock concentrations or grazing at inappropriate times during the year simply because forage is available can quickly return the riparian area to its preproject level of deterioration.

Monitoring. The only way to ensure that the grazing management strategy is effective in meeting its objectives is to monitor the results. One of the best, and simplest, ways of doing this is to establish photo points of a few representative areas. This should be done not only for the bottomlands, but for key upland sites as well (Chaney et al. 1993). Photographs should be supplemented with observations of the condition and trend of riparian vegetation, as well as streamflow and streambank characteristics. Definitive and logical changes to grazing management strategies can be made when monitoring results are compared from one year to the next.

Revegetating Riparian Areas
Damaged by Livestock

Revegetation should be performed only after site characteristics, such as water availability and channel stability, are taken into consideration. Ecosystem managers should not underestimate the resiliency of riparian ecosystems and the potential for native riparian plant recovery once the causes of decline are addressed. Understanding this potential may prevent the use of revegetation in areas that are fully capable of recovering on their own. The experience at Sheepshead Spring (case study 2) underscores this point; also see chapter 4.

Livestock and riparian revegetation projects are usually incompatible. Even relatively small numbers of livestock can significantly influence planting results. Hall (pers. comm., 1990), for example, noted that considerable damage was done to a revegetation site near Burro Creek, in northeastern Arizona, when one cow managed to gain entrance through an open gate. Plantings that are palatable to livestock (e.g., cottonwood and willow seedlings) should be protected during the first two growing seasons (Hall, Forbis, Bell, Goodwin, pers. comm., 1990). Although nonpalatable species may not be subject to grazing, they should nevertheless be protected from indirect grazing impacts, such as trampling.

Of course, the most effective way to protect newly planted vegetation is to keep livestock out of the revegetation site altogether. Several authors have summarized various methods for excluding livestock, including barriers, stream crossings, and water access points (Reichard 1984; Platts and Nelson 1985a). Maintenance and monitoring are important no matter what strategy is used to protect planted vegetation. As exemplified by the revegetation results at Burro Creek, keeping gates closed and preserving fence integrity are as important as fence installation.

Two imaginative planting strategies have successfully established vegetation despite continued livestock pressure. Bammann and Sims (pers. comm., 1990) used long cottonwood and willow poles (more than 3.5 m long) that were less vulnerable to damage from trampling and grazing. This strategy, however, requires the auguring of deep holes, which may limit its applicability to areas where site characteristics (e.g., site accessibility, alluvial characteristics, geology) do not prohibit drilling. Planting nonpalatable species around palatable species may also be worthy of future consideration as a planting strategy in the face of continued grazing pressure (Goodwin, pers. comm., 1990).

Recreation

Recreational activities such as hiking, camping, and driving off-road vehicles (e.g., motorcycles, dune buggies) can impact the ecological condition of riparian areas (Manning 1979). Out of the 27 riparian restoration projects evaluated by Briggs (1992), recreation was considered to be at least partly responsible for deterioration in the ecological condition of two sites and influencing recovery efforts at two others.

Ecosystem managers should be aware of the potential effects of recreation on riparian ecosystems and how recreational activities can affect the results of recovery efforts. Several authors review strategies for assessing the environmental impacts of recreation. Settergren (1977) discusses three strategies for measuring recreational impacts, and Magil and Twiss (1965) discuss using photographs to assess aesthetic and biological changes.

Recreational Impacts

Recreational activities can modify habitat by compacting soils and increasing soil erosion (Lutz 1945; Manning 1979), which in turn can produce changes in the structure, density, and composition of vegetation (Hammitt and Cole 1987).

Hiking and Related Activities

Hiking in general, and trampling in particular (e.g., around camping sites or frequently visited areas), can negatively affect soils by removing leaf litter and other organic matter on the soil surface, compacting the soil and reducing its macroporosity, and increasing soil erosion (Willard and Marr 1970; Manning 1979). Impacts can occur even when recreation pressures are light. Cole (1995), for example, found that a single night of camping can significantly reduce vegetation height and cover. In general, sites that experience heavy recreational use are more likely to be characterized by soils that have less organic matter on the surface (Dotzenko et al. 1967), greater compaction (Lutz 1945), reduced water infiltration rates (Frissell and Duncan 1965; Settergren and Cole 1970), and less depth (Merriam and Smith 1974) than the same soils in areas that do not experience significant recreational use.

Changes in soil characteristics can indirectly influence plant vigor. Soil compaction may decrease available nutrients even in areas with relatively high levels of organic matter and soil nutrients (Young and Gilmore 1976). In addition, Hopkins and Patrick (1969) observed that increased soil compaction may decrease or even prevent root extension, possibly decreasing a plant's ability to acquire adequate amounts of water. Settergren and Cole (1970) found that the amount of water available to plants in the soil surface was significantly reduced in areas that experienced heavy recreational use.

Increased soil erosion caused by human trampling may also lead to root exposure (Merriam and Smith 1974). Exposed tree roots tend to be more vulnerable to drought and mechanical injury and may increase the vulnerability of trees to wind (Manning 1979).

Off-Road Vehicle Use

Off-road vehicles, such as motorcycles and dune buggies, can cause considerable damage to the environment in localized areas. Although off-road vehicles tend to affect the condition of natural ecosystems in much the same way as hiking, the impacts tend to occur much more quickly. Webb (1983) noted, for example, that motorcycles can produce noticeable changes to the environment even after just one pass.

Off-road vehicles compress the soil, increase soil bulk density, and alter vegetation cover (Snyder et al. 1976; Webb 1983). These changes can significantly alter sediment and water runoff characteristics. In a study that compared undisturbed areas to similar areas that were disturbed by off-road vehicle use, the total unit runoff (volume per unit area) from the disturbed area was nearly eight times more than from the undisturbed area (Snyder et al. 1976).

Addressing Degradation Caused by
Recreational Impacts

Ideally, riparian ecosystems should be managed so that recreational activities can occur with minimal impact on the integrity of the ecosystem. The key for management is to adjust use to a level where the system does not become impaired (Johnson and Carothers 1982). Reducing public access is a practical strategy — however, although severely restricting or prohibiting recreation may be tempting to ecosystem managers, such strict regulation probably should not be implemented for the long term because it ignores the value of these areas to people. Moore (1989) noted that in between the extremes of no management and restriction, there are a variety of visitor management techniques that can be used to adjust the behavior of recreationists so that their impact on riparian ecosystems is minimized. These techniques include the following:

Keeping trails and campsites away from vegetation communities that are most vulnerable to recreational impacts. Results from trampling experiments suggest that shrub-dominated vegetation types are more resistant to damage than those dominated by erect forbs (Cole 1993). In addition, some vegetation types are damaged quickly by trampling but also recover quickly (e.g., subalpine herbaceous vegetation; Cole 1995). Guidelines that describe which plant species are most resistant to recreational impacts under varying site conditions can be helpful to ecosystem managers who are planning to use revegetation to improve the condition of degraded riparian ecosystems. Unfortunately, only preliminary work has been done in this area: see Liddle 1975, Magil and Leiser 1967, Merriam and Smith 1974, and Wager 1966.

Confining camping to a small number of campsites instead of dispersing use across a large number of sites. The rationale behind this strategy is that many vegetation types are affected by use frequencies as minimal as one night per year, yet the extent of damage does not always increase in proportion to use frequencies (Cole 1995). This strategy will keep the damage confined to a relatively small area.

Providing signs that explain the conservation values of staying on a trail or within a camping site.

Identifying sensitive areas and resting them until they recover, or eliminating use in these areas altogether (Craig 1977).

Promoting natural recovery of damaged areas by aerating the soil and adding fertilizer (Post 1979).

Revegetating Areas That Experience Recreational Use

It can be difficult to revegetate riparian areas that experience frequent human use. The results at several revegetation sites evaluated by Briggs (1992) were affected by human activities. Planted cottonwood and willow poles were collected for firewood, seedlings were run over by off-road vehicles, and irrigation systems were sabotaged. Given these experiences, ecosystem managers should consider postponing revegetation until recreational use is reduced or eliminated (at least in the short term).

This approach may be more effective than simply planting vegetation because it addresses the causes (or at least one of the causes) of site deterioration, possibly allowing the site to recovery naturally. For example, Richter and Richter (pers. comm., 1990) noted that curtailing off-road vehicle use and eliminating a trailer park were two factors that led to the dramatic natural recovery of riparian vegetation at the Nature Conservancy's Hassayampa River Preserve in Arizona.

If recreational use cannot be reduced, there are some practical strategies that ecosystem managers can employ to prevent damage from plantings. For example, Sims and Goodwin (pers. comm., 1990) noted that damage from some human activities is often unintentional and can be avoided by posting signs that briefly explain the objectives and importance of the revegetation project. Fencing off the revegetation site or temporarily closing hiking trails that pass near the site also can be used to lessen impacts.

Competition from Nonnative Species

Nonnative plant species have invaded and successfully established along many drainages in the southwestern United States. Thickets of saltcedar (*Tamarix* spp.) and Russian-olive *(Elaegnus angustifolia)*, for example, are now common along many rivers in Arizona and New Mexico, and Bermuda grass *(Cynodon dactylon)* and red brome *(Bromus rubens)* are abundant in many low-lying areas. Although nonnative species can negatively affect the establishment and growth of native species, the presence of significant numbers of nonnative species in a riparian area is usually a symptom rather than the cause of riparian decline; nonnative species appear to outcompete native species only after the riparian environment has been affected by other factors. Saltcedar, for example, has replaced many native species partly because it is better adapted to the artificial flow regimes created by impoundment (Warren and Turner 1975). Directly addressing this cause of riparian decline by altering impoundment flow (or removing the dams) to favor native species may reverse the decline yet may be impossible in the short term due to our perceived dependency on controlling natural river flow.

To be successful, riparian recovery projects need to address these other factors. However, in some situations ecosystem managers may have to limit their efforts to addressing the symptoms rather than the causes of riparian decline. Removing saltcedar and revegetating with native species has been successful in reestablishing native riparian species in locations where their ability to establish naturally has been greatly reduced (Briggs et al. 1994). Anderson (pers. comm., 1993) noted that revegetating riparian areas choked with saltcedar often produces limited results, yet this may be one of the few options remaining to reestablish native seed sources.

Revegetating When Exotics Are Present

Dense growth of on-site vegetation will compete with artificially planted vegetation for moisture, nutrients, sunlight, and space, possibly to the extent of influencing establishment rates. Anderson and Ohmart (1985) found a significant negative correlation in both survival and biomass production between vegetation planted in areas with Bermuda grass and in areas where Bermuda grass was either absent or removed prior to planting. They found Bermuda grass roots wrapped around the root systems of newly planted trees at depths of 2.4 m and surmised that the Bermuda grass probably competed with the planted trees for scarce water and nutrients.

In addition, some exotics may attract pests that can also affect the condition of native riparian species. Southern pocket gophers *(Thomomys umbrinus)* and the larvae of Apache cicadas *(Diceroprocta apache),* both of which are attracted to Bermuda grass, can damage newly planted vegetation by eating roots or increasing the trees' vulnerability to disease (Anderson and Ohmart 1982). Accordingly, depending on the density and type of vegetation, ecosystem managers should consider removing undesired species to give planted vegetation an opportunity to establish.

Nonnative Vegetation Control

Revegetation projects in areas characterized by a high density of nonnative species will have to allocate resources for clearing vegetation prior to planting and for controlling nonnative regrowth after planting has been completed. Such actions can greatly improve the effectiveness of revegetation but should be planned carefully so that plants valuable to wildlife are not destroyed.

Anderson (pers. comm., 1992) noted that weed control should be planned for at least the first growing season following planting. He observed that lack of weed control has led to mortality rates exceeding 80% in three years on a few of his revegetation sites along the Colorado and Kern Rivers.

Saltcedar is an aggressive colonizer that has become established in dense thickets along many parts of the Gila (Marks 1950; Turner 1974) and Colorado Rivers (Robinson 1965) since the 1940s. The high rate and longer period of seed production by saltcedar and its ability to withstand relatively long periods of inundation make it difficult to eradicate and allow it to outcompete native riparian species in many areas (Warren and Turner 1975; Ohmart et al. 1977).

Successfully controlling saltcedar can be the key to revegetating areas choked by this aggressive nonnative species (Barrows 1993). Hall and Moore (pers. comm., 1990) noted that clearing saltcedar prior to planting improves establishment rates of planted species. However, saltcedar can come back more aggressively following initial clearing. The results of several revegetation projects demonstrate that clearing saltcedar prior to planting provides planted seedlings with a head start, allowing artificially planted riparian vegetation to establish successfully despite prolific saltcedar regrowth (Briggs 1992).

There are many strategies for controlling saltcedar. Anderson (pers. comm., 1990) clears saltcedar with a bulldozer before planting. He cautioned that any type of clearing should be done carefully so that valuable wildlife habitat is left intact (e.g., snags). Such drastic methods as using a bulldozer should be considered only in extreme situations. Much of Anderson's revegetation work is in areas where saltcedar has completely choked out native species (e.g., the lower Colorado River).

Controlling saltcedar with direct application of herbicide can also be effective. Cutting saltcedar to a few centimeters above the ground and applying herbicide directly to the stump has been successful (Richter and Richter, pers. comm., 1990; Sudbrock 1993). The effectiveness of this technique is greatly improved when the following guidelines, compiled from observations by Neill (1990), Richter and Richter (pers. comm., 1990), and Sudbrock (1993), are taken into consideration: (1) saltcedar removal (in some circles known as a "Tammi-wacking party") should be done in the late fall when sap is flowing downward and seed dispersal is not bolstered by removal activities; (2) trees should be cut to within 5 cm of the ground surface; (3) the entire surface of the cambium layer should be cut and treated; (4) herbicide (Tordon RTU, Tordon 101, Garlon 3A, or Garlon 4) should be applied within minutes after trees have been cut (Sudbrock [1993] noted that Garlon 4 can be diluted one-to-one with either diesel oil or water); (5) sites should be revisited within one year of the original treatment so that resprouting foliage can be cut and treated.

Treatments done improperly usually fail and may even increase the density of saltcedar (Neill 1990). In addition, mismanagement of toxic chemicals could have far-reaching impacts on local flora and fauna and could contaminate groundwater. Neill (1990) also discusses the cost, the efficiency, and treatment rates of the above-mentioned herbicides, as well as the tools that are most effective in cutting saltcedar. Howard et al. (1983) noted that Garlon 4 can be effective during autumn months on stumps 10 to 15 cm above the ground.

Wildlife

Riparian Revegetation and Wildlife

Wildlife rarely contribute significantly to the decline of riparian ecosystems. However, some species of wildlife can affect the results of revegetation. The riparian revegetation experiences in Arizona revealed that elk *(Cervus ela-phus)*, white-tailed deer *(Odocoileus virginianus)*, and beaver *(Castor cana-densis)* are capable of eating significant quantities of planted vegetation; and rice rats *(Oryzomys palustrus)*, woodrats (*Neotoma* spp.), and pocket gophers (*Thomomys* spp.) can chew through irrigation lines (Briggs 1992). The impacts of these and other wildlife species should therefore be taken into account when ecosystem managers are developing the revegetation design (fig. 3.2).

Protecting Planted Vegetation from Wildlife

If revegetation is used, ecosystem managers should assess the potential for damage from these species of wildlife. If the potential for damage is high, ecosystem managers should consider planting in alternative locations, implementing techniques to lessen the extent of damage (e.g., install cages around planted vegetation to prevent beaver damage), or planting additional propagules to compensate for potential loss from wildlife.

A study of a group of riparian revegetation projects in Arizona indicates that beaver frequently affected the results of revegetation (Briggs 1992) by removing pole plantings for dam construction and devouring aboveground growth to the point where seedlings could not survive (fig. 3.2). Ecosystem managers should be especially wary of planting in beaver-populated areas that do not have alternative sources of woody, riparian vegetation in addition to species that are going to be planted.

Trapping beaver and moving them off-site requires large expenditures of time and labor and is usually only marginally successful in preventing beaver from influencing revegetation results. Furthermore, removing all beaver from a site may not be beneficial. Brayton (1984) and Braasch and Tanner (1989) noted that in some situations beaver dams can positively influence recovery

Figure 3.2 Cottonwood tree damaged by beaver along Burro Creek, northeastern Arizona. (Photograph by Robert Hall, Bureau of Land Management)

by constructing dams that slow water flows, thereby lessening streambank erosion.

Beaver cages (chicken wire wrapped around the base of the tree) were used in several revegetation projects to protect planted trees from beaver damage (Briggs 1992). Hall and Forbis (pers. comm., 1990) noted that beaver cages were generally successful in protecting planted vegetation from beaver damage as long as water levels did not frequently inundate the site to a depth where beaver could swim up and consume the vegetation growth above the level of the cage.

At least in the southwestern United States, beaver appear to be the principal wildlife species that affects the results of naturally watered riparian revegetation projects. Other wildlife species become more of a problem when irrigation is used: during dry months, irrigation lines are a tempting source of water for all kinds of wildlife species. Pollock (pers. comm., 1990) noted that he constantly had to repair irrigation lines that had been gnawed through by woodrats and pocket gophers. Burying irrigation lines may help, yet the increased labor that this requires may make this option unacceptable. Wildlife damage to irrigation lines is yet another factor that should be considered before using irrigation to establish vegetation in riparian ecosystems.

4 / Natural Recovery in Riparian Ecosystems

Natural recovery is one of the ecosystem manager's strongest allies, and promoting it by addressing the causes of degradation should be the aim of all riparian recovery efforts. Fortunately, partly due to the high levels of natural disturbance that riparian ecosystems typically experience, riparian ecosystems are generally highly resilient and have considerable capacity for recovery. An evaluation of a group of riparian revegetation efforts in Arizona revealed that previously damaged riparian sites experienced dramatic natural regrowth of vegetation in at least two different scenarios: following significant flooding and after the causes of site decline were addressed (Briggs et al. 1994).

At Aravaipa Canyon, Arizona (case study 3), for example, natural regrowth following a significant flood event (approaching a 500-year event) was so extensive that the results of revegetation work performed following the flooding could not be located. The results in Aravaipa Canyon are not unique. All told, 32% of the successful riparian revegetation sites evaluated in Arizona experienced prolific natural regrowth that completely obscured the results of artificial planting efforts (Briggs et al. 1994). In other words, these projects achieved their objectives primarily because of natural regenerative processes, not artificial plantings.

Dramatic natural regrowth of riparian vegetation has also been documented after the bottomland environment has been relieved of a major disturbance. For example, Case (1995) reported a 106% and 800% increase in the mean height and crown volume, respectively, of black cottonwood *(Populus trichocarpa)* along Meadow Creek (a tributary of the Grande Ronde River in northeastern Oregon) after the cessation of grazing.

The results of these and other experiences indicate the importance of delaying active recovery efforts until a better understanding is obtained of how the ecosystem will respond to the removal of disturbance factors. In other words, ecosystem managers need to recognize the potential for dramatic natural regrowth following flooding and after the causes of site decline have been addressed and should consider the possibility that artificial revegetation may not be necessary.

Case Study 3
Aravaipa Creek
Natural Recovery after Flooding

Aravaipa Creek flows through Aravaipa Canyon, draining portions of the Galiuros and Pinaleño Mountains in southwestern Arizona. In October 1983, a 500-year flood passed through Aravaipa Canyon, removing significant amounts of streamside vegetation (fig. 4.1). The damage was considered so severe that many thought it would take generations for the streamside vegetation communities to recover.

To encourage the recovery of Aravaipa's riparian communities, a large revegetation project was completed a year after this destructive flood event. More than 2,000 cottonwood *(Populus fremontii),*

Figure 4.1   Aravaipa Canyon, Arizona as it appeared less than one year after it experienced flooding that exceeded magnitudes of a 500-year event. (Photograph by Albert R. Bammann, Bureau of Land Management)

Figure 4.2 The same view as in figure 4.1 eight years after flooding. All the regrowth here was the result of natural process, not artificial plantings. (Photograph by Albert R. Bammann, Bureau of Land Management)

Goodding willow *(Salix gooddingii),* and Arizona walnut *(Juglans major)* propagules were planted. At the time of the site evaluation seven years after the flood damage had occurred, natural regrowth was so extensive that the results of the artificial revegetation effort could not be found (fig. 4.2).

These experiences also underscore the need to consider more than just the vegetative characteristics of a bottomland area when assessing its health and the need for implementing a recovery effort. A riparian ecosystem may be "healthy" even if it is characterized by low vegetation density, diversity, and volume. As the experience at Aravaipa Canyon illustrates, large flood events can remove vast amounts of streamside vegetation, and yet the consequences of flooding often produce sites that have characteristics ideal for the rapid return of many of the same species that were removed.

This chapter provides the ecosystem manager with background information that will assist in predicting the potential that a site will experience strong natural regrowth. The overall goal is to provide ecosystem managers with a predictive tool that will help them to avoid wasting recovery projects in areas that are fully capable of coming back on their own. Given the numerous factors that affect natural recovery (proximity to seed sources, seed characteristics, alluvial deposition, competition from other plants, moisture availability, etc.), the best that ecosystem managers will be able to do is make a general prediction regarding the potential for natural recovery (e.g., it is likely that significant natural recovery will occur during the next three years).

Factors Influencing Natural Recovery

Of course, the discussion on natural recovery needs to take into consideration not only the likelihood for seed germination, but also the likelihood that the species will actually establish. For example, it is common to find "carpets" of cottonwood seedlings during early to midsummer along many low-elevation drainages in the Southwest (fig. 4.3). However, only a small percentage of these seedlings will survive through the first year, and an even smaller percentage (maybe none) will survive to maturity. The vast majority of the seedlings will be removed by subsequent flooding or perish as a result of the effects of numerous other factors.

Predicting when and where natural regrowth will take place is complicated by numerous factors that affect the natural distribution and propagation of streamside vegetation species. These factors include spatial and temporal variation in the "seed bank" (Brokaw 1985), variation in scour and deposition (Gecy and Wilson 1990), inundation and floodwater dispersion (Finlayson et al. 1990), elevation (Ericsson and Schimpf 1986; Szaro 1990), drainage area, geology, and flow regime (Zimmerman 1969), among many others. Of these factors, a few appear to have an overriding role in determining the extent, location, and timing of natural regrowth. Of the revegetation sites evaluated

Figure 4.3 A "carpet" of cottonwood seedlings along the lower Verde River near Phoenix, Arizona. (Photograph by Liz Rosan, Sonoran Institute)

by Briggs (1992), those that experienced prolific natural regeneration had three critical characteristics in common.

First, many of the sites had also experienced floods of sufficient magnitude to remove streamside plants, deposit fresh alluvium, and saturate floodplain soils—characteristics vital for germination and seedling establishment (Fenner et al. 1985; Stromberg et al. 1991). Second, seed sources were located in or near all of these riparian sites. This is an elementary principle: seed sources need to be available for nonvegetative natural regeneration. (However, determining whether or not seeds are "available" is complicated.) Third, the riparian ecosystem was not characterized by any of the major factors that often cause the ecological condition of riparian ecosystems to decline, namely, low groundwater, channel instability, high soil salinity, and severe direct impacts. Those four factors also have an important role in determining the effectiveness of riparian revegetation efforts. This, of course, is not surprising. The likelihood that a plant will establish and survive is determined by many of the same factors whether it is placed in the ground naturally or artificially.

Postponing revegetation even for one season can provide clues that may help to better predict the magnitude of natural regrowth, leading to a more accurate estimate of the potential effectiveness of revegetation. Given the recuperative powers of riparian ecosystems in conjunction with large flood events, postponing revegetation for sites that recently (during the last two years) experienced flooding may be particularly advisable. Postponing revegetation at Aravaipa Creek just two years would have probably revealed clues of the extensive natural regrowth that was going to occur at that site.

Seed Availability

When an ecosystem manager is determining the potential for natural regrowth in riparian ecosystems, the question that needs to be posed is "Are seed sources of appropriate species available?" If seed sources of desired species are unable to influence the target site, artificial revegetation may be the only method of reintroducing desired species. This question can be difficult to answer. Does "available" mean that seed sources need to be in the immediate degraded riparian site, or can seed travel to and germinate in distant locations?

In some cases, answering the seed availability question will be relatively straightforward. For example, the answer is obvious when desired species are in the immediate site or if the reverse is true, when the closest seed sources of the desired species are found only in locations very distant from the site (case study 4). Most situations will probably fall in between these two extremes.

Determining whether seed sources of desired species are too far removed to influence the degraded riparian site requires stepping back from the target site to take into consideration vegetation characteristics along upstream and downstream reaches and tributaries. For example, if cottonwoods are the desired species, then the potential that neighboring cottonwood stands (if present) will be able to provide seed to the site needs to be assessed. Making such an assessment, however, can be challenging, requiring an understanding of the autecology of each of the desired species, particularly the seed dissemination characteristics.

Unfortunately, basic research is lacking regarding the distances that even the most common woody riparian species can disperse seed. There is some anecdotal information indicating that seeds disseminated by some members of the willow family (e.g., Fremont cottonwood and Goodding willow) are capable of reaching sites 1–2 km downstream, indicating that such wind-dispersed species may be able to provide viable seed to relatively distant locations. Observations by Forbis, Hall, and York (pers. comm., 1990) as well as information included in studies by Farmer and Bonner (1967), Everitt (1968), Glinski (1977), White (1979), Strahan (1981), Fenner et al. (1984), McBride and Strahan (1984), Reichenbacher (1984), and Brady et al. (1985) indicate that seed disseminated by many riparian species can travel to distant sites.

The amount of viable seed produced, seed size, length of dissemination period, and mode of dissemination (e.g., streamflow, wind, animals, and birds) all influence the distance that seed can be dispersed (Scott, pers. comm., 1992). Hardin's (1984) study of Eastern cottonwood *(Populus deltoides)* suggests that the age and density of trees in a given area may also influence seed dissemination. His results indicate that younger trees produce greater quantities of smaller, lighter seeds than older trees. Seeds of older trees are therefore likely to be carried shorter distances by wind compared to seeds of younger trees. This may negatively influence the ability of an older tree to disseminate seeds by air to distant locations, especially if the tree is in an area removed from the main channel, where streamflow is unlikely to aid in seed dispersal. Hardin (1984) also observed that seed production of isolated trees may be reduced by lower pollen availability.

The relative number of seeds disseminated also parallels seed size, with light-seeded species (e.g., cottonwood, willow, sycamore, ironwood) producing an abundance of seeds and somewhat heavier-seeded species (e.g., Arizona walnut) producing fewer seeds (Streng et al. 1989). The distance that seeds are dispersed generally declines with increasing seed size. However, flooding can extend the dispersal range of many riparian species (as well as allow foreign species to come into an area). The distance that seeds can travel with streamflow depends on many factors, including the length of time seeds remain viable, flow rates, deposition patterns, and seed buoyancy characteristics. Skoglund (1990) noted that although the number of seeds in drift material is biased toward species with a long floating time, a few species with nonbuoyant seeds may be subject to water dispersal.

Factors encouraging good seed availability include the following: light, wind-dispersed seeds are within 2 km of the damaged site; heavier seeds, particularly those that are animal dispersed, are within about 0.5 km; the seed source is composed of numerous healthy trees of seed-producing age; and the damaged site is located downstream from the seed source and is on an active floodplain, allowing seed transport to the target site to be aided by streamflow. Factors contributing to poor seed availability include the following: heavier seeds are more than 0.5 km from the target site; the seed source is composed of isolated trees; and the target site is removed from the active floodplain.

Implementing a strategy that will promote the germination of seed intro-

McEuen Seep is a small, isolated spring near Fort Thomas, Arizona, that was revegetated following years of overuse by livestock. McEuen Seep is an oasis that is surrounded for miles on all sides by the comparatively austere Sonoran Desert environment. Prior to its decline, this spring was characterized by perennial aboveground flow and a lush riparian vegetation community that consisted of overstory species such as Arizona walnut *(Juglans major)* and Fremont cottonwood *(Populus fremontii)*. At the time of revegetation, none of these phreatophytes remained.

The main objective of revegetation was to reestablish the historical species and, by so doing, improve habitat for wildlife (this site is considered extremely important for many species of wildlife, particularly migratory birds). Another objective was to reduce accelerated erosion rates that had produced gullies, which in several locations had penetrated to bedrock.

Of the revegetation projects evaluated in Arizona, the McEuen Seep project probably exhibits the most effective use of revegetation. Site conditions indicated that the target site could support the desired species (i.e., water availability, soil salinity, and channel stability were not considered severe enough to prevent establishment of these species) as long as the causes of decline were addressed. In addition, the isolated location of the site made natural regeneration of species such as cottonwood and walnut unlikely, making revegetation an even more appropriate mitigation option.

Livestock were excluded from this area and several overstory species were planted, including Fremont cottonwood, Arizona sycamore *(Platanus wrightii),* and Arizona walnut; deer grass *(Muhlenbergia rigens)* was also planted. Thirteen years later, the condition of McEuen Seep has improved remarkably. Some of the planted overstory species are more than 15 m tall, and herbaceous understory species have also reestablished (fig. 4.4). In addition, numerous woody seedlings in the understory provide evidence that planted species are now contributing to natural establishment (Robles, pers. comm., 1995).

Figure 4.4   McEuen Seep near Fort Thomas, Arizona, is a good example of an effective use of revegetation. (Photograph by Liz Rosan, Sonoran Institute)

duced naturally to the damaged riparian site is one of the principal reasons for trying to determine natural seed dispersal characteristics of nearby desired species. In a study along Boulder Creek, Colorado (case study 5), for example, investigators encouraged the establishment of native woody species from natural seedfall through site manipulations that mimicked characteristics vital to the species' germination and growth (moist, devegetated areas). The Boulder Creek work provides a model for future riparian recovery efforts because it embodies the idea of improving site conditions through methods that work with and augment natural processes.

To this point, the discussion has focused only on propagation via seed dissemination. Vegetative reproduction, however, is a common postdisturbance revegetation mechanism for many riparian plant species (Gecy and Wilson 1990). Arizona sycamore *(Platanus wrightii)* (Glinski 1977), burrobrush *(Hymenoclea monogyra)* (Campbell and Green 1968), and seepwillow *(Baccharis salicifolia)* (Stromberg et al. 1993) are examples of woody riparian species that commonly reproduce by sprouting from stems, lateral roots, or trunk bases. Vegetative reproduction is also a common form of regeneration for many herbaceous species in low-elevation bottomland ecosystems (Stromberg et al. 1993). For some species, therefore, flooding can produce the conditions required for natural regrowth, even if it does not occur when seeds are being produced.

Flooding

Many riparian species are adapted to, and depend on, flooding for propagation. Vegetation patterns along streams and washes often reflect this dependence, with streamside vegetation communities often forming a mosaic composed of vegetation of different ages, varying colonizing abilities, and tolerance of flooding and shade (White 1979). For example, woody riparian vegetation communities often exhibit a well-defined age diversity, ranging from young seedlings found mainly along the stream side of typical point bars to well-established woodlands on the higher elevations (Everitt 1968; McBride and Strahan 1984). Flood disturbances play an important role in the initial establishment of riparian vegetation (Everitt 1968; Reichenbacher 1984; Hupp and Osterkamp 1985; Campbell and Green 1986), revitalizing riparian ecosystems by producing conditions that can lead to a rapid return to predisturbance conditions (Stromberg et al. 1993).

High flow events can produce sites that initially lack competition from other plant species, are exposed to full sunlight, and are characterized by denuded coarse soils with high moisture availability (White 1979). Such characteristics are perfect for the germination and establishment of cottonwoods and willows (Everitt 1968; Stromberg et al. 1991), as long as newly established roots can follow retreating water levels (Fenner et al. 1984; Reichenbacher 1984).

Whether a particular flood event will produce the conditions necessary for the establishment of vegetation depends to a large extent on its magnitude and timing. On one hand, a large-magnitude flood will be likely to deposit alluvium in areas less vulnerable to the scouring effects of frequent, low-magnitude flow events. On the other hand, a low-magnitude flood (e.g., of less than a two-year return interval) may not have the energy to scour channel banks and reduce understory and overstory competition.

Since floods also play a critical role in seed dispersal, the timing of a flood event is another major factor to consider when determining the potential for natural regrowth. In the Southwest, many riparian species produce seed during the time of year when high flows are likely. Fremont cottonwood *(Populus fremontii)*, Goodding willow *(Salix gooddingii)*, Arizona walnut *(Juglans major)*, Arizona sycamore, and velvet ash *(Fraxinus velutina)* disperse seed between May and June, months when high flows are common as a result of snowmelt and spring rains. By contrast, mesquite (*Prosopis* spp.) is adapted to germinate in summer or fall—well after the spring floods—when temperatures are high (Mooney et al. 1977) and high flows during what is the monsoon season in the southwestern United States help to distribute and scarify mesquite seed.

Optimal seedling recruitment conditions occur when a high-magnitude flow event (e.g., a ten-year return interval or greater) takes place during the time of year when seed is being disseminated. If this occurs, there is a good chance that the flow event will carry seed to downstream sites and deposit it there. In addition, the flood event will moisten depositional surfaces at the appropriate time, providing emerging seedings with the moisture required for growing roots to permanent water.

Autecology of Selected Southwestern Riparian Tree Species

Determining the potential for natural regrowth requires a basic understanding of the autecology of the various species being considered for the revegetation project. It is beyond the scope of this book to review the autecological characteristics of even a small percentage of southwestern riparian species; the four

Case Study 5
Boulder Creek
Promoting Natural Recovery Using
Existing Seed Sources

Boulder Creek is a perennial stream near Boulder, Colorado, that flows east from the Indian Peaks of the Rocky Mountains to Saint Vrain Creek, a tributary of the South Platte River. As with riparian forests in many parts of the western United States, those along Boulder Creek are failing to reproduce, as a result of upstream river impoundment. The principal objective of this effort was to gain a better understanding of how plains cottonwood *(Populus deltoides)* and peachleaf willow *(Salix amygdaloides)* can be established in areas where channel migration and flood disturbance no longer occur. Specifically, several methods were tested that artificially created some of the major effects that floods have on streamside ecosystems (Friedman et al. 1995).

Streamside sites were disturbed using a track-mounted excavator. An average of 16.5 cm of sod was removed from the soil surface to create areas free of competing vegetation (fig. 4.5). Synchronizing the removal of sod with the peak of cottonwood and willow seed release ensured the influx of a large number of willow and cottonwood seeds relative to those of other species. In addition, moisture availability was enhanced by using sprinklers, which delivered on average 0.98 cm of water each day. Although seeds of plains cottonwood and peachleaf willow were collected locally and sprinkled onto several of the study plots, those that were seeded by natural seedfall experienced similar germination rates. Cottonwood densities of 10.3 seedlings per square meter were achieved using these methods (Friedman et al. 1995). Peachleaf willow establishment was much less, apparently because daily irrigation was insufficient to prevent drought stress during the first month.

Experiences at this project demonstrate how some riparian species, particularly those that produce dependable crops of seeds capable of germination, can be reestablished in areas where natural establishment is being prevented, even when seed sources are scattered and not abundant. Using local trees as the source for seed also helps to maintain the genetic composition of the site, avoiding potential problems that may result when off-site plant materials are used (Scott, pers. comm., 1995).

Figure 4.5 These sites near Boulder Creek, Colorado, were cleared of vegetation to encourage the regrowth of plains cottonwood and peachleaf willow from natural seedfall. (Photograph by Michael Scott, U.S. Fish and Wildlife Service)

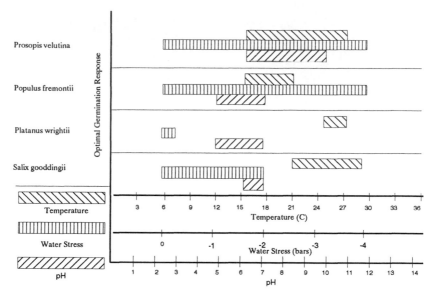

Figure 4.6 Optimal germination responses of four southwestern riparian species to temperature, water stress, and pH. (Illustration by Mark Briggs)

that are discussed simply provide examples of the type of information that will help in determining the potential for natural regrowth, as well as the information required to match planting materials to site conditions conducive to establishment and growth (fig. 4.6). An excellent autecology reference that will provide additional, and much more detailed, information on this subject is Young and Young 1992; another example of a localized reference work that would be valuable to an ecosystem manager is Kearney and Peebles 1969, which includes taxonomic descriptions of all the bottomland vegetation families and genera found in Arizona and the means for identifying them.

Fremont Cottonwood

Of the major southwestern riparian tree species, the autecology of Fremont cottonwood *(Populus fremontii)* is probably the best documented. The timing of seed dispersal, seed physiological characteristics, and seed germination and seedling growth characteristics all indicate a species adapted to the highly dynamic floodplain environment (Farmer and Bonner 1967; Stromberg 1993).

Fremont cottonwoods are dioecious trees, producing fragile seeds that are approximately 1 mm in length. A tuft of trichomes catches air currents, helping to disperse the seeds (Fenner et al. 1985). Seeds are released during the spring, and viable seed tends to germinate within 48 hours after being dispersed. Siegel and Brock (1990) found the best germination response was between 16°C and 21°C. Fremont cottonwood seeds tend to have a short life span, losing viability within five weeks of dispersal (Fenner et al. 1984). Natural recovery that occurred following the completion of several riparian recovery projects in Arizona seemed to indicate that Fremont cottonwoods are capable of providing viable seed to areas 1–2 km distant. These results, however, were realized following large flood events that were capable of moving seed long distances quickly.

For cottonwood seedlings to survive, moist conditions must persist until roots are able to obtain more consistently available subsurface water. Root growth rates of young seedlings average 6 mm per day (Fenner et al. 1984).

Arizona Sycamore

Arizona sycamore *(Platanus wrightii)* is a monoecious tree that flowers in late spring, producing ripe seeds during October of the same year that fall intact over the next several months (Bock and Bock 1985). Fowells (1965) noted that *P. occidentalis* (a related species found in southeastern Europe and the Himalayas) begins producing seed when it reaches 25 years of age, but optimum seed production occurs when the tree is between 50 and 200 years old. Bock and Bock (1985) found that each fruit contained an average of 667 seeds and each tree dispersed an average of 449 seeds per square meter, with 90% of the seeds falling between 10 m upstream and 16 m downstream. These trees also reproduce asexually by stump or basal sprouting. Bock and Bock (1985) found trees growing in clumps from buried branches. Regardless of the means of reproduction, seedling survival depends greatly on two factors: the persistence of moist conditions until the onset of summer rains, and protection from scouring floods.

Germination of Arizona sycamore seed appears to be very sensitive to temperature, water stress, and pH. Siegel and Brock (1990) found significantly better germination at 27°C with 0 bars water stress. Germination rates fell from more than 50% at 0 bars water stress to 20% at −2 bars water stress. There is a general germination reduction of about 5% for every two weeks of seed age (Siegel and Brock 1990).

Goodding Willow

True willows (*Salix* spp.) are found in southwestern riparian ecosystems in association with other riparian species such as cottonwood, velvet ash, and Arizona walnut. Siegel and Brock (1990) found that Goodding willow *(Salix gooddingii)* seed germinated best at 27°C. Germination decreased significantly at −2 bars water stress. Results of some riparian recovery efforts indicate that Goodding willows are capable of providing viable seed to areas 1–2 km downstream (Briggs 1992). As with Fremont cottonwoods, results were realized following large flood events that were capable of moving seed large distances quickly.

Velvet Mesquite

In comparison to the species mentioned above, mesquites (*Prosopis* spp.) are better adapted to establishing and growing on landscapes farther removed from drainageways, where depth to saturated soils is significant. However, only along river floodplains or the edges of lakes do the trees grow to a large size and occur in dense woodlands (Hennessy et al. 1985; Warren and Anderson 1985). The ability of mesquite to survive in dry conditions is due, in part, to a tap root that can reach deep subsurface water (in contrast, species such as *Salix* spp. tend to have a more shallow, lateral root system). Mesquite roots often grow to depths of 15 m (Campbell and Green 1986).

In contrast to cottonwoods and willows, mesquites are adapted to germinate in summer or fall, when temperatures are high (Mooney et al. 1977) and high flows during what is the monsoon season in the southwestern United States help to distribute and scarify mesquite seed. Velvet mesquite *(Prosopis velutina)* seeds appear to have a wide range of acceptable conditions for germination; they germinate well between 16°C and 27°C, and 0 and −4 bars water stress (Siegel and Brock 1990). Simpson (1977) reported that seeds of the mesquite genus retain viability for years. Such long-term viability can impact the spatial and temporal dynamics of the floodplain seed bank, which may have important implications for changing patterns of canopy replacement after disturbance (Streng et al. 1989). A review of the ecology of, threats to, and recovery potential of mesquite forests in Arizona is provided in Stromberg 1992.

5/ Water Availability

Many riparian plants can establish and survive only in areas where they can develop root systems to saturated soils (Campbell and Green 1968; Fenner et al. 1984; Reichenbacher 1984); in addition, their root systems are relatively shallow and spread out great distances laterally but frequently do not penetrate more than 3 m below the soil surface. The combination of these two characteristics makes it very difficult for these species to survive if the zone of saturation drops significantly at the riparian site (Fenner et al. 1984; McBride and Strahan 1984; Busch et al. 1992). Low groundwater levels also place severe limitations on the feasibility of reestablishing riparian plants (Swenson and Mullins 1985; York and Hall, pers. comm., 1990; Briggs et al. 1994). This point is underscored in the experiences at a riparian revegetation site in Arizona called Box Bar, where only those plants placed in areas of relatively high water availability survived (case study 6).

In the Southwest, groundwater decline is considered a major factor in the deterioration and disappearance of several major riparian areas. The San Pedro River, one of the Southwest's last natural, low-desert rivers, is being threatened by increased groundwater pumping by the nearby town of Sierra Vista (Davis 1995). Groundwater decline due to pumping (McNatt et al. 1980; Stromberg et al. 1992) and flow manipulation caused by Horseshoe Dam and Bartlett Dam (Forbis, pers. comm., 1990) have degraded cottonwood stands along the lower Verde River, Arizona. Dramatic changes to the structure and diversity of riparian ecosystems along the lower Colorado River are due in part to changes in water availability (Ohmart et al. 1977). Increased groundwater pumping in the Santa Cruz River basin of Arizona is at least partly to blame for degrading cottonwood stands along the Santa Cruz River (Stromberg et al. 1993). A formerly extensive mesquite bosque in Casa Grande Ruins National Monument, Arizona, has been almost completely eliminated by groundwater decline due to overpumping of groundwater (Judd et al. 1971).

Riparian ecosystems tend to respond to water stress in hierarchical fashion (Taub 1987). Low levels of stress affect ecosystems at the level of the individual. For example, diversion of surface water may cause some trees to suffer

Case Study 6
Box Bar
Lessons in the Importance of Water Availability

Box Bar is located on the lower Verde River roughly 5 km south of Bartlett Reservoir, just north of the outlying reaches of Arizona's largest city, Phoenix. The results of this revegetation project offer a good example of how difficult it can be to establish native, phreatophytic species in areas with low-lying groundwater.

The objective of this revegetation project was to improve habitat for wildlife and to gain a better understanding of the feasibility of establishing Fremont cottonwood *(Populus fremontii)* and Goodding willow *(Salix gooddingii)* in a degraded riparian ecosystem. The deterioration of cottonwood stands along the lower Verde River is blamed on a combination of factors, including groundwater withdrawals by the city of Phoenix, manipulation of natural river flow by Bartlett and Horseshoe Dams, and drought (McNatt et al. 1980; Forbis and Pollock, pers. comm., 1990).

The Box Bar site consists of an upper floodplain and a secondary channel. The upper floodplain, which lies between the main Verde River channel and the secondary channel, is 2.5–3 m higher in elevation than the secondary channel. According to project personnel, groundwater levels on the upper floodplain frequently drop to more than 3 m below the soil surface.

In 1979, a revegetation effort was undertaken to reestablish riparian trees on the upper floodplain. More than 600 Fremont cottonwood and Goodding willow poles and seedlings were planted in this effort. Because additional plantings were available, 7 poles were planted in the secondary channel. Those planted on the upper terrace received irrigation for the first three summers (seven days a week, two hours per day).

Eleven years later, only the trees that were planted in the secondary channel appear healthy and show strong potential that they will survive in the long term (Fenner, pers. comm., 1995). These cottonwoods are now approximately 10 m in height (fig. 5.1) and are in marked contrast to the two cottonwoods that survived on the upper floodplain—both of which show obvious signs of water deprivation (stunted growth, yellow leaves, and sparse foliage).

Figure 5.1 At a revegetation site called Box Bar along the lower Verde River, Arizona, only those seedlings planted in the secondary channel survived. Despite extensive irrigation, those planted on the upper terrace (on the right) did not survive. (Photograph by Liz Rosan, Sonoran Institute)

physiological stress that results in reduced growth rates and increased mortality (Medina 1990; Smith et al. 1991). Higher levels of water stress can cause community-level and ecosystem changes such as reductions in biomass and elimination of species.

How Groundwater Decline Occurs

Any phenomenon that alters the pressure on groundwater will cause the level of groundwater to change. These phenomena include differences between the supply and withdrawal of groundwater, meteorological and tidal influences, urbanization, earthquakes, and external loads (Todd 1980).

Groundwater levels can vary over a period of years. These long-term changes in groundwater levels can be produced by alternating series of wet and dry years, when precipitation is above or below average. However, recharge, not precipitation, is the factor that governs (assuming constant annual withdrawals) groundwater levels. Recharge depends on rainfall intensity and distribution, and surface runoff (Todd 1980). A pronounced downward trend in groundwater levels will be found in areas where overdraft consistently exceeds recharge (such as in highly urbanized basins).

Streamflow arising from groundwater discharge is called base flow. Streamflow and groundwater levels generally show seasonal patterns (Bock and Bock 1984). During periods of frequent precipitation, streamflow mainly originates from surface flow, whereas during extended dry periods all streamflow may be contributed by base flow (Hall 1968). Streamflow variations are closely related to groundwater conditions where a stream channel is in direct contact with an unconfined aquifer. Depending on the relative level of the zone of saturation, the stream may recharge the groundwater or receive discharge from the groundwater. A stream that is receiving groundwater discharge is a gaining stream; a stream that is recharging groundwater is a losing stream (fig. 5.2). The rate of loss or gain within the channel is controlled by channel permeability which, in turn, is a function of the texture of the substrate (Gebhardt et al. 1989).

Flow characteristics can change significantly as a stream descends from upland areas to the desert floor. In comparison to streams in mesic climates that are often gaining streams and have increased discharge per stream length, streams in arid and semiarid climates are often losing streams and tend to have decreased discharge per stream length. For example, in less than 15 km, flow characteristics along Rincon Creek, Arizona (case study 1; see chapter 2),

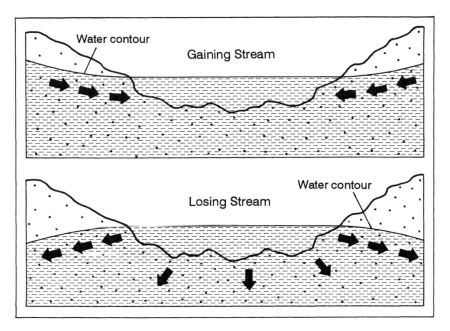

Figure 5.2 Comparison of water contours and groundwater flow directions for gaining and losing streams. (Illustration by Mark Briggs)

change from perennial along its uppermost reaches, to intermittent along its mid reaches, to ephemeral along its lower reaches. Along this same distance, maximum annual depth to saturated soils changes from less than 0.5 m along uppermost reaches to more than 16 m along the lowermost reaches.

Determining groundwater conditions for riparian ecosystems along ephemeral and intermittent streams is not always an obvious matter. This may be particularly true for bottomland environments near urban and agricultural settings where groundwater pumping can change groundwater conditions quickly. When high-capacity wells went into operation along Tanque Verde Creek, Arizona, for example, measured rates of groundwater decline were 3.7–6.5 m per year near the pumps (where cones of depression are deepest) and 0.5–1.5 m per year for areas farther removed from the pumping stations (Stromberg et al. 1992).

An area characterized by ephemeral streamflow and rapid changes in groundwater levels can provide ecosystem managers with few obvious indications of the current status of groundwater. Even remnant phreatophytes in a damaged riparian ecosystem should not be taken as evidence of stable groundwater conditions. Mature phreatophytes with well-developed root systems may persist where newly planted vegetation cannot. To develop an effective revegetation plan, ecosystem managers must investigate current groundwater conditions, how they have changed, and how they may be affected in the future.

Evaluating Groundwater Conditions

Comparing past to present groundwater conditions can provide ecosystem managers with a good sense of groundwater stability, and how and to what extent groundwater characteristics have changed. Such a comparison will help ecosystem managers develop realistic recovery objectives and strategies, as well as provide important information for predicting the potential for natural recovery (see chapter 4).

Understanding Past Groundwater Conditions

Groundwater depths and seasonal-fluctuation characteristics have been determined for many areas and can be obtained from a variety of sources. Past hydrogeologic studies of the degraded riparian environment can be extremely helpful, providing not only up-to-date information, but also descriptions of past groundwater conditions, explanations of the changes that have occurred and the reasons behind the changes, and a review of additional sources of information. The U.S. Geological Survey collects and publishes extensive data on groundwater conditions for even the most remote parts of this country. In addition, some universities have hydrology departments that can provide a great deal of information regarding the types of data available and the studies that have been completed in the area of interest.

Wells that have been monitored for an extended period of time can provide valuable information concerning current and past groundwater conditions. State and county offices responsible for water resources data are good places to begin searches for well records; many cities have offices with similar functions. In addition, private well owners often keep detailed well records that describe changes in groundwater depth. However, these records should be used with caution. For different reasons, at least a portion of the well records may not provide information that depicts the actual groundwater level in the riparian zone; ecosystem managers should pursue gaining access to measure water levels directly. Well levels may have been inaccurately recorded, or some of the wells may have been drilled into perched or confined aquifers, providing a confusing picture of the groundwater conditions. Including numerous wells in the survey will allow managers to weed out erroneous data, improving the chances of obtaining an accurate picture of groundwater characteristics.

Old photographs can provide valuable hydrologic information. An old photograph of a large cottonwood gallery forest, for example, not only reveals information about the aboveground vegetation characteristics, but also indirectly provides information about groundwater. For a cottonwood forest to have existed, the groundwater level must have been close to the surface. Photographs can be obtained from historical societies, local residents, and libraries. Photograph repositories (aerial and ground) for the U.S. Geological Survey are listed in the bibliography.

Determining Present Groundwater Conditions

There is no substitute for performing your own field investigations. Collecting data directly is often more credible than obtaining data from other sources. It also allows ecosystem managers to tailor investigations to answer the questions necessary for designing a riparian revegetation plan.

An investigation of current groundwater conditions should include depth to groundwater, groundwater quality, how depth to groundwater varies within the riparian site, and how these conditions vary throughout the year. An initial objective of such an investigation should be to develop a map that accurately depicts the current elevation of the top of the zone of saturation in the riparian area. Ideally, the groundwater map should also depict how the elevation of the zone varies during the year. Such information will provide ecosystem managers with the data needed to confidently choose vegetation species, plan irrigation schedules, and develop planting designs tailored to the specific area, greatly improving the overall effectiveness of a revegetation project.

Although vegetation characteristics may reflect current groundwater conditions, such data can be misleading if not combined with direct measurements. As mentioned previously, remnant mature species are able to obtain water from greater soil depths than can seedlings, and their presence, therefore, may not mean that newly planted vegetation of the same species can survive. Also, surviving species may have tapped into a perched water table or some other localized source of water that may not be available to seedlings.

Personnel and monetary constraints may limit the thoroughness of the groundwater investigations. Concentrating groundwater investigations during the time of the year when water availability is likely to be the lowest is an alternative that will provide ecosystem managers with information describing the most severe conditions newly planted vegetation will have to endure. In the Southwest, the most severe time of year occurs during early summer, prior to the onset of the monsoon season. During this time of the year, lack of precipitation and cloudless, hot days create a situation when water availability is at its lowest and the water demands of many plant species are at their highest.

Using Local Wells to Monitor Depths to Saturated Soils

A well on or adjacent to the proposed revegetation site is an obvious prize that ecosystem managers should take full advantage of. Wells can be common in even relatively remote locations (e.g., windmill-operated pumps for cattle), and ecosystem managers should explore the revegetation site and bordering areas to see if any exist.

Measuring the water depth is usually a straightforward process that can be accomplished in a variety of ways. Lowering a steel tape or a weighted tag line into a well is one method that is both simple and accurate. Adding chalk to the measuring device produces an obvious submersion line, providing the distance from the top of the well to the water surface.

Obviously, the more wells that are measured, the more accurately the depth to the zone of saturation can be represented. However, even monitoring the water depth in only one well can provide valuable information. A map of the riparian site depicting the elevation of the top of the zone of saturation can be developed by accounting for surface elevation differences between points of known groundwater depth (e.g., wells) and points of unknown depth; the accuracy of the map will become increasingly questionable with increased surface distance from the monitoring well.

Installing Wells

Gathering groundwater information is critical to developing the recovery plan. If no wells are available on site, then installing wells to monitor groundwater conditions may be a viable option. This option will become increasingly desirable if the following factors hold true: first, if the site's geohydrologic and vegetation characteristics offer few clues as to the current depth of groundwater; and second, if the cost of installing a well (determined to a large extent by the type of well, drilling depth, and soil characteristics) would not put the recovery effort over budget.

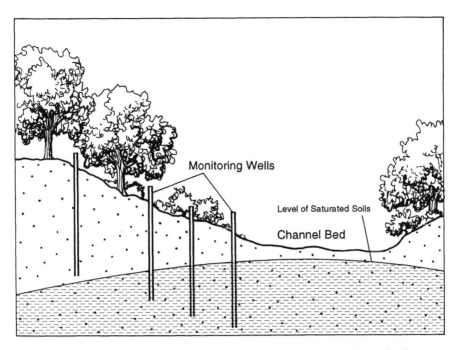

Figure 5.3 Location of monitoring wells across an idealized stream channel. (Illustration by David Fischer)

In comparison to developing wells that furnish potable water, installing wells for monitoring groundwater depth can often be done much more inexpensively using perforated PVC pipes. The number and locations of monitoring points required to give a realistic picture of groundwater depth and fluctuation depend on several factors, including the size, topography, geology, and soil characteristics of the site. Small sites characterized by relatively homogeneous hydrogeologic conditions will require fewer monitoring points than larger, more complex sites (Dalton et al. 1991). Installing a line of wells perpendicular to the stream channel will provide information on how groundwater levels change with distance from the main channel (fig. 5.3).

Monitoring Frequency

Monitoring groundwater levels even for a single year can provide ecosystem managers with a rough gauge of how the depth to the zone of saturation can fluctuate. Obviously, the more often the depth to groundwater is measured and the longer the monitoring period, the more accurately the average groundwater fluctuations for the riparian site can be described. Knowledge of how groundwater conditions vary throughout the year is important because large seasonal groundwater fluctuations can significantly affect the results of revegetation. Large groundwater fluctuations can drown planted vegetation during some parts of the year and leave them high and dry during other parts of the year (this is particularly important if planted vegetation is left without water during the hot and dry summer months).

Constraints may limit the number of times that the depth to water is measured. If only a few measurements per year are possible, then measuring during the time of year when depths to groundwater are likely to be the greatest will provide an indication of the most severe conditions that planted vegetation is likely to experience.

Revegetating Riparian Ecosystems Characterized by Groundwater Decline

Planting riparian plants at a site of low water availability (defined by where groundwater levels frequently fall 3 m or more beneath the surface) is difficult. Artificially planted phreatophytes will not survive if their root systems cannot reach the zone of saturation. More than 80% of the unsuccessful revegetation projects evaluated by Briggs (1992) experienced low survival rates of planted vegetation (0% in some cases) due to low water availability (case study 6, the Box Bar, is an example of this).

Addressing the causes of groundwater decline is often difficult because the factors causing the decline often affect large areas and may be beyond the jurisdiction of ecosystem managers. Such factors include groundwater pumping, river impoundment, agricultural irrigation, and even climatic change. In such situations, ecosystem managers will be faced with developing a plan for improving the condition of a degraded riparian area where groundwater decline will continue to be a problem.

Although groundwater decline may be the most significant cause of degradation, other factors also may play a part in degrading the ecological condition of the site. Identifying and then addressing these other causes should also be a the priority. If the negative impacts of these other causes can be reduced, there is a much greater chance that the condition of the degraded site will improve, even if groundwater decline continues to be a major problem (see chapter 3).

Matching Plant Materials to Site Conditions

If revegetation is going to be used, making certain that the planting materials are adapted to site conditions is an obvious step that can greatly improve the chances for establishment. Matching the water needs of the planting materials to the site's water availability is key in this regard. However, it is important that ecosystem managers realize the possibility that the site's hydrologic conditions may have changed so dramatically that planting riparian species may no longer be practical. If this is the case, the overall goal of the recovery project may need to be changed from restoration (trying to re-create the ecosystem that existed prior to significant human impacts) to rehabilitation.

At several of the riparian revegetation sites evaluated by Briggs (1992), for example, species such as cottonwoods and willows were planted in areas where the depth to groundwater was significant. In such instances, planting species better adapted to the drier conditions may have been a sound alternative even though the created vegetation community would have been different than the "original." For example, revegetating the damaged reach of Rincon Creek (case study 1; see chapter 2) with such overstory species as mesquite (*Prosopis* spp.), palo verde (*Cercidium* spp.), and netleaf hackberry *(Celtis reticulata)* — species that are native to the area and capable of surviving outside of the immediate floodplain — may be a practical alternative to trying to reestablish Arizona walnut *(Juglans major)*.

Planting Techniques

Due to their smaller tissue mass and less-developed root system, seedlings are especially vulnerable to low water availability. This is a particular concern in arid climates where moist alluvium deposited during spring flooding dries rapidly with the onset of extreme summer temperatures. Therefore, most riparian plants can survive only if the depth to the zone of saturation increases slowly enough to provide developing root systems sufficient time to grow down through the soil to more permanent water.

Some planting techniques have produced good results when used in sites characterized by relatively deep groundwater (where the zone of saturation typically exceeds a depth of 3 m). These techniques are described below. Ecosystem managers need to recognize, however, that these techniques have limitations and should be used either on a small scale (to determine their effectiveness before large amounts of money and time are wasted) or in combination with non-phreatophytes and additional recovery strategies (e.g., land management changes, bank stabilization structures).

Using Large Poles

York (pers. comm., 1990) and Swenson and Mullins (1985) have successfully planted tall Fremont cottonwood and Goodding willow poles (in some cases more than 7 m long) in areas where groundwater is more than 3 m below the soil surface. This technique, however, has at least two limitations. First, mechanized equipment is often necessary either to auger holes to relatively deep groundwater or to punch poles through the soil. Consequently, the site must be accessible to mechanized equipment. Second, it can be difficult, if not impossible, to drill or dig deep holes in riparian areas characterized by coarse alluvium containing high compositions of cobble and rock. In such situations, revegetation may have to be abandoned or changed to include plant material capable of establishing in areas characterized by low-lying groundwater.

Drilling to Groundwater

Anderson (pers. comm., 1990) has successfully planted phreatophytes in riparian sites characterized by relatively deep groundwater. He accomplished this by using an auger (18 cm in diameter) to drill to the zone of saturation, refilling the holes with displaced alluvium or mulch, planting seedlings on top of the refilled holes, and supplying the seedlings with irrigation water. The drilling action of the auger breaks up compacted soil and clay lenses, allowing irrigation water to flow to the zone of saturation with less resistance. Developing roots will follow the moisture gradient to groundwater, often producing taproot development in species like cottonwood and willow that typically have shallow lateral root systems.

Digging to saturated soils can be accomplished by drilling with a motorized auger or digging with a backhoe, or with a variety of manual instruments. Holes should be backfilled so that air spaces, which appear to be detrimental to developing vegetation, are kept to a minimum. This technique is restricted by some of the same factors that limit where large poles can be planted. The consistency of overlying alluvium can prevent drilling, and site location may prevent access by heavy mechanized equipment.

The cost of augering to groundwater (about $2 per hole when many holes are being drilled) is justified by the increase in growth and survival of planted species, decreased irrigation requirements, and reduced labor costs (Anderson et al. 1984). Trees planted after augering to groundwater were significantly taller, had greater total growth and foliage volume, and experienced lower mortality rates than trees planted in areas that were not tilled (Anderson and Laymon 1988).

Irrigating

A well-planned irrigation system can provide newly planted vegetation with sufficient moisture to allow roots to grow to saturated soils. However, the irrigation of plantings will not always produce high establishment rates, and the cost may be prohibitive as well. The Box Bar revegetation project (case study 6) illustrates the inherent difficulties associated with promoting the growth of phreatophyte roots to relatively deep groundwater. Of the revegetation projects evaluated in an Arizona study, only six used irrigation and, out of that number, only one experienced a survival rate of planted vegetation greater than 20%.

The irrigation systems used in several of these projects appeared to be capable of wetting only 1–2 m^3 of the vertical soil profile, which is often insufficient for sites where the groundwater drops to 2–3 m below the soil surface. In some cases root development occurred only within the wet profile, and vegetation was left high and dry when irrigation was stopped (Goldner 1981; Hall and Forbis, pers. comm., 1990).

In addition, even the most efficient irrigation systems require frequent attention to repair and refuel pumps; patch, replace, and realign pipes; and unclog emitters. This can be especially problematic when a riparian revegetation project is located in a remote area.

As mentioned earlier, Anderson (pers. comm., 1990) has developed revegetation strategies that combine augering with irrigation. He makes two recommendations based on his revegetation experiences. First, plantings should be irrigated for at least the first growing season. Irrigation should be stopped only after the plantings have developed roots to the zone of saturation. The length of the irrigation period depends largely on how long it takes the vegetation to develop roots to the zone of saturation. In most cases this can be discerned by observing tree growth: trees tend to grow faster and leaves appear greener after the tree roots contact groundwater. Second, irrigate during the summer for 90 working days (5 days on, 2 days off) at a rate of 8 gallons/tree/day (30 liters/tree/day).

Carothers et al. (1990) noted that many planted trees are able to reach groundwater 3 m below the soil surface when irrigated for two seasons after

Figure 5.4 The results of this revegetation effort demonstrate the importance that depth to saturated soils can have on survival rates of artificially planted vegetation. The poles in the foreground did not survive even though they were planted on a landscape that is less than a meter higher in elevation than the successful plantings in the background. (Photograph by Liz Rosan, Sonoran Institute)

having been put into the ground. They also noted that the most reasonable irrigation strategy is to give planted vegetation an overabundance of water, so that the soil is saturated to groundwater nearly constantly.

Plant Placement Strategies

Distance to groundwater can vary considerably, even between planting sites separated only by several meters (fig. 5.4). Strategically placing planting materials to take advantage of landscape characteristics that offer greater water availability can significantly affect survival rates. Secondary channels, depressions, and other low-elevation sites where water collects and where distance to groundwater is minimal are obvious candidates. This strategy, however, embodies one of the principal difficulties inherent in revegetating bottomland environments. Vegetation planted in low-elevation sites is often prone to flood damage. Planting designs often walk a thin line between placing vegetation in areas where it may be removed by high-frequency flow events (e.g., ten-year flood) and planting in areas that offer protection from flooding yet are characterized by low water availability. This point is pursued in more detail in chapter 6.

Revegetating Areas That Experience Prolonged Inundation

The discussion throughout this chapter has focused on the problems associated with revegetating riparian ecosystems with low water availability. On the other side of the coin are areas that are inundated for extended periods of time. Most species found in riparian ecosystems can tolerate periodic flooding without damage because of special metabolic and physical features such as adventitious roots, porous cell structure, and specific metabolic pathways (Walters et al. 1980).

Despite these adaptations, phreatophytes will be damaged if inundated for an extended period of time. Areas that are characterized by fine-textured soils and high water availability may create an anaerobic environment that surrounds the root systems of planted vegetation. An anaerobic environment stresses a variety of plant processes, including water and nutrient uptake, xylem and phloem transport, photosynthesis, and transpiration. The short- and long-term impacts to plant health from growth in an anaerobic environment are well documented. For example, Burrows and Carr (1969) noted that the roots of many riparian species either become dormant or begin to die when they are flooded. A plant's tolerance to flooding is primarily determined by its ability to grow adventitious roots and new secondary roots under low oxygen conditions (Teskey and Hinckley 1977).

Newly planted vegetation is more susceptible to damage from inundation than are mature plants, and prolonged inundation can damage vegetation planted in bottomland environments, affecting revegetation results (Briggs 1992). For example, cottonwood poles planted along the lower Verde River (Pollock, pers. comm., 1990) and cottonwood and willow poles planted along the Rio Grande south of Albuquerque, New Mexico (Swenson and Mullins 1985), did not survive three weeks of inundation.

Ecosystem managers therefore have to be extremely careful when choosing planting sites. Again, monitoring groundwater conditions prior to planting will provide the information needed to define areas that are characterized by shallow groundwater yet are not prone to lengthy periods of inundation. If a revegetation site does experience frequent and prolonged inundation, ecosystem managers should choose plant species that are more tolerant of inundation. Walters et al. (1980) and Teskey and Hinckley (1977) classified tree tolerance to flooding as very tolerant, tolerant, intermediately tolerant, and intolerant, depending on the condition of the root system following various durations of inundation:

Very Tolerant. Tree species that can maintain their "normal" roots in an at least partly anaerobic rhizosphere while also producing new secondary and adventitious roots. These species can withstand flooding for periods of two or more growing seasons.

> *Alnus oblongifolia,* New Mexico alder
> *Alnus tenuifolia,* thin-leaf alder
> *Baccharis salicifolia,* seepwillow
> *Phragmites communis* var. *berlandieri,* reed
> *Platanus wrightii,* Arizona sycamore
> *Populus angustifolia,* narrow-leaf cottonwood
> *Populus fremontii,* Fremont cottonwood
> *Populus tremuloides,* quaking aspen
> *Salix exigua,* coyote willow
> *Salix gooddingii,* Goodding willow

Tolerant. Tree species in which the normal root system deteriorates but the plant can produce adventitious roots to replace it. These species can withstand flooding for most of one growing season.

Acer negundo, boxelder
Atriplex canescens, fourwing saltbush
Atriplex confertifolia, shadscale
Atriplex lentiformis, lens-scale, quail bush
Celtis reticulata, netleaf hackberry
Fraxinus velutina, velvet ash
Juglans major, Arizona walnut

Intermediately Tolerant. Tree species that will produce few new roots and are usually dormant during periods of inundation. These species can survive flooding for periods between one and three months during the growing season.

Cercidium floridum, blue palo verde
Chilopsis linearis, desert willow
Lonicera involucrata, inkberry
Lycium spp., wolfberry
Prosopis glandulosa, honey mesquite
Prosopis velutina, velvet mesquite
Prosopis pubescens, screwbean mesquite
Robinia neomexicana, New Mexico locust

6/ The Drainageway

Human history is replete with examples of river "improvement" projects that were designed with an attitude that often ignored Mother Nature's intimations. Dams alter natural streamflow patterns (often to the detriment of streamside vegetation communities), streambank stabilization structures have all too often led to increased channel instability, and numerous riparian revegetation projects have donated carefully prepared planting materials to streamflow simply because project designers did not take into consideration channel dynamics and geomorphologic processes.

This chapter focuses on evaluating the stability of alluvial stream channels. An alluvial channel is partly or fully bounded by sediment derived from prevailing conditions of water-sediment discharge, and this sediment is subject to reworking by streamflow. Unlike bedrock channels, which are confined by rock outcrops and change slowly over a long period of time (hundreds of years), self-adjusting alluvial channels are prone to changes over much shorter time periods. A single large erosive flow, for example, can dramatically change channel width, potentially affecting the results of revegetation and other recovery efforts.

Abrupt changes in channel characteristics — even the destruction of streamside vegetation by erosive flows — are not necessarily unnatural, nor are they necessarily undesirable. Particularly in arid and semiarid climates where precipitation can be flashy (sudden, erratic, dramatic, and of short duration) and streambanks are often sparsely vegetated, alluvial channels are susceptible to extensive and sometimes dramatic channel erosion. Although such disturbances often remove large amounts of streamside vegetation, they also produce conditions that make natural recovery likely, as in Aravaipa Creek, Arizona (case study 3; see chapter 4).

Such dynamics make any type of streamside recovery effort risky. However, a sound understanding of stream processes will allow ecosystem managers to

The development of this chapter is, in large part, due to the generous technical guidance and editorial comments of Waite Osterkamp (hydrologist, U.S. Geological Survey).

predict the direction, type, and magnitude of channel adjustment, improving the chances that recovery strategies will work in harmony with natural stream processes, rather than against them (Leopold 1977; Heede 1981; Keller and Brookes 1984; Schumm et al. 1984; Brookes 1985; Harvey and Watson 1986).

Some channel reaches, such as the Babocomari River, which flows through Huachuca City, Arizona, are inherently unstable, greatly reducing the chances that riparian revegetation or other improvement strategies will be successful (case study 7). Such channels may no longer be experiencing equal rates of deposition and erosion and have fallen out of what is known as dynamic equilibrium. As compared to relatively stable alluvial channels, those that have fallen out of dynamic equilibrium are more likely to alter their geometric form over a short interval of geomorphic time (Leopold and Maddock 1953; Wallace and Lane 1976; Schumm et al. 1984).

Channel Instability and Riparian Ecosystems

The abrupt erosional and depositional changes that often accompany channel instability can damage property and infrastructure, remove bottomland vegetation, and devastate riparian revegetation projects. Loss of riparian vegetation due to increased floodplain instability has been documented along Dry Creek (McBride and Strahan 1984), the Carmel River (Kondolf and Curry 1984), and the lower San Lorenzo River (Griggs 1984) in California, the San Pedro River (Zimmerman 1969), Sonoita Creek (Glinski 1977), and the Gila River (Minckley and Clark 1984) in Arizona, and the Rio Puerco in New Mexico (Elliott 1979).

These studies show that channel instability can degrade associated riparian ecosystems in two general ways. First, streamside vegetation can be destroyed directly through bank erosion and the resultant loss of channel bank and floodplain integrity. Second, abrupt channel erosion can dewater the riparian zone, often increasing the depth to which phreatophyte roots must grow to reach saturated soils.

Any land use that alters soil and water runoff can trigger instability and initiate fluvial adjustment along an entire drainage system. Overgrazing by livestock, plowing, wagon and cattle trails, road construction, and urbanization (Elliott 1979; Schumm et al. 1984); or any of a number of upland watershed activities and overt up- and downstream channel alterations, including channelization, dams, and bank training work (Heede 1981); or even changes in climatic factors (Leopold 1951) can alter watershed runoff and erosion

Case Study 7
Babocomari River
Revegetating in an Unstable Stream Channel

A revegetation project along a reach of the Babocomari River near Huachuca City in southeastern Arizona epitomizes the difficulty of revegetating unstable stream channels. This reach of the Babocomari River has experienced dramatic changes in channel characteristics, possibly due to land use changes in the watershed that have altered the sediment and water discharge characteristics (e.g., livestock grazing and urbanization). Erosive flows have lowered the elevation of the channel bed, making the channel deep and narrow. In several parts of this reach, streambanks have become unstable, and continuing erosion threatens the integrity of Huachuca City's sewage treatment pond, which lies roughly 50 m from the river channel.

The objective of this project was to stabilize streambanks and prevent the channel from eroding into the sewage pond. Reestablishing vegetation against the bank was considered key to stabilizing the banks in this area. However, incised channel conditions reduced the options available for planting. Vegetation had to be planted either on the abandoned floodplain, where depth to groundwater was extreme, or in the channel, where planting materials could be removed by minor flooding. Since groundwater decline and other factors ruled out planting in the floodplain, planting materials were placed in the active channel.

Extreme measures were required to prevent plantings from being removed by flow events. In this case, a rail and wire fence was installed in the channel along the streambank that lies closest to the sewage pond. The fence was more than 100 m long and was constructed with steel poles 10 cm in diameter placed in concrete-lined holes 2.5 m apart (fig. 6.1).

One hundred eighty dormant Goodding willow *(Salix gooddingii)* poles measuring 2.5 m long and 10–15 cm in diameter were planted 1.5 m apart in holes 1.5 m deep, parallel to the fence, between the fence and the eroded banks of the river. The holes were dug using motorized augers. After planting, a drip irrigation system was installed. (Also, following planting, the tops of the poles were painted with an oil-base paint to prevent sunburn and transpiration loss.)

Despite the extravagant and relatively expensive methods that were used in this project, only about 10% of the plantings have

Figure 6.1 An in-stream rail and chain-link fence installed along the Babocomari River near Huachuca City, Arizona, to promote deposition along the toes of eroded banks; *Salix gooddingii* poles were planted behind the instream structure. (Photograph by Mark Briggs)

survived. However, accumulated debris next to the fence has helped to protect the streambank from the direct impacts of streamflow. This project demonstrates some of the obstacles inherent in revegetating near alluvial stream channels that have fallen out of dynamic equilibrium and have experienced dramatic changes in channel characteristics.

characteristics. These alterations can initiate periods of enhanced channel and floodplain change (Wallace and Lane 1976; Heede 1981). Depending on the size of the watershed, years may pass before the effects of a perturbation in one part of the watershed are felt in the bottomlands. This means that channel reaches not characterized by obvious signs of instability at the time the recovery project is implemented can exhibit instability years later as a result of undetected disturbances in other parts of the drainage system.

Channel Dynamics

What Is Dynamic Equilibrium?

The concept of geomorphic dynamic equilibrium was first used by Gilbert (1880) in his discussion of landform changes in an entire drainage basin, not just the stream channel. Based on this early work, the dynamic equilibrium concept was applied to channels that show a long-term balance between erosion and deposition (Mackin 1948; Leopold and Maddock 1953; Hack 1960).

The underlying premise of the dynamic equilibrium concept is that the characteristics of an alluvial channel adjust to water (Q) and sediment (Q_s) discharge. A given reach of an alluvial channel can be seen as an energy regime where a delicate balance exists between its geometric form and the inflow and outflow of energy (Bradford and Priest 1980). Dynamic equilibrium conditions exist when the channel size, shape, sediment characteristics (roughness), and gradient are constant through time, and the delivery rates of all particle sizes of sediment to the channel (or channel reach) are equal to the rates of these sediment sizes leaving the channel reach (Osterkamp and Harrold 1982). A stream reach at dynamic equilibrium is stable in that it tends to maintain its form and local gradient over a period of years (Leopold et al. 1964; Van Haveren and Jackson 1986).

This does not mean that stable channels do not experience changes in channel characteristics. Due to continual (though not necessarily constant) changes in independent variables — mainly sediment load and discharge — channel characteristics are changing continuously, and stream reaches will make incremental or periodic adjustments in channel geometry (including bank cutting and cycles of streambed scour and fill) (Leopold 1964; Hendrickson and Minckley 1985; Van Haveren and Jackson 1986).

At dynamic equilibrium, the energy of flow is such that coarse sediment sizes supplied from upland areas, as well as suspended sediment, can move through the channel without causing bank erosion (Osterkamp and Harrold

1982). An alluvial channel reacts to changes to sediment and water discharge by altering several variables that describe its channel geometry and sediment characteristics. The relation of the primary variables (channel width, mean depth, gradient, roughness, median particle size of channel material, and silt-clay content of channel material) to water and sediment discharge is summarized by the equation $[W, D, G, n, d_{50}, SC, \ldots] = f(Q, Q_s)$. This relation includes other dependent variables (e.g., turbulence), but those listed here appear to be the most important.

Major Channel Adjustment Processes

Once an alluvial channel becomes unstable, it will initiate a variety of adjustment processes that will reestablish a state of dynamic equilibrium. Channels adjust to changes in sediment and water discharge in several different ways. Some channel adjustment processes are subtle, requiring relatively little energy, whereas others require more energy and are obvious. Channel adjustments involve changes in bed form, bed armor, width, pattern, and longitudinal profile (Heede 1980).

Bed Form Changes. Bed form is associated with the topography of the channel bed. Bed form changes as a result of the interaction of streamflow patterns and bed materials. Typically, bed form changes in a progressive downstream manner, from a poorly defined pool-step structure in the headwaters, to better developed pool-riffle sequences in gravel bed reaches, and finally to sand bed forms in downstream reaches. This generality does not hold for stream channels outside the Basin and Range province of the southwestern United States.

Bed forms contribute to the resistance to flow associated with bed roughness, and their development generally requires the least amount of energy and time (Simons and Richardson 1966). Streambed form self-adjusts to regulate resistance to flow in a manner that effectively dissipates energy over a wide range of flow and bed material conditions (Morisawa 1985). Bed form adjustment following a change in soil and water runoff characteristics (e.g., formation of a gravel bar) may be sufficient to restore dynamic equilibrium.

Channel Width Adjustments. Periods of channel widening are often short-lived, occurring during erosive flow and often facilitated by the destruction of streamside vegetation or other stabilizing properties. Channel widening can occur during a single flood, or it can occur in stages. As a channel widens, its mean depth tends to decrease, the channel straightens and gradient increases, the median particle size of bed material generally increases, and the bedload discharge of the stream increases after the flood (Osterkamp and Harrold 1982). Increase in channel width expands the wetted perimeter of the flow, leading to greater roughness of flow and lesser flow velocities.

During nonerosive flow, alluvial channels tend to narrow or change in a manner that promotes narrowing (Osterkamp and Harrold 1982). The rate that a channel narrows is dependent greatly on the particle-size distribution of sediment that is delivered to it. An alluvial channel will narrow much more rapidly when the sediment delivered to it has an abundance of fines, rather than coarse sediment grains. Following a channel-widening flow, an extended period of time may be required for a channel to narrow when available sediment from upland areas is deficient in fine sizes.

Channel Pattern Changes. As a stream widens, it can break through meanders, reducing its sinuosity and stream length, frequently leading to a change in channel pattern (Heede 1980). The pattern of a given channel can change briefly or for an extended period of time. In addition, all three channel patterns (braided, meandering, and straight) are feasible along the same drainageway at a given time.

The reasons why a stream changes from one pattern to another are not precisely understood. Investigations into this issue have shown that the complete range of channel patterns is dependent on stream power, which, in turn, reflects sediment load and discharge (Chorley et al. 1984). Therefore, one influence on channel pattern is the energy of flowing water as it transports water and sediment through the channel. Channel pattern is also affected by tributary influences. Tributaries can introduce different types of sediment load into the channel, potentially causing changes in channel sinuosity and pattern. Changes in slope (as a result of large quantities of sediment introduced by tributaries, for example) can also cause changes in channel pattern. As the slope increases, channel sinuosity may also increase to maintain a constant gradient at a given location (Chorley et al. 1984).

The pattern of alluvial channels is also influenced by sediment load. Osterkamp (1978) observed, for example, that large amounts of sediment (relative to stream competence) must be available for transport for braiding to occur. The resulting aggradation can cause the depth of flow to become so shallow that it can no longer occur in a single channel. The channel system then divides into several narrow channels to increase depth of flow and sediment-carrying capacity (Shen and Vedula 1969).

The pattern and shape of a channel are influenced by several processes,

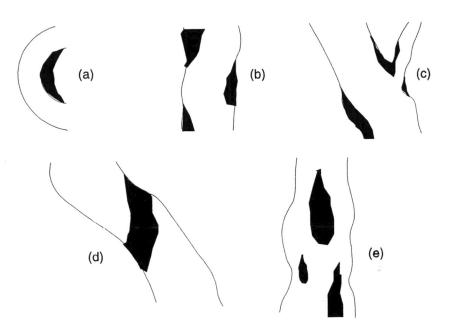

Figure 6.2 Examples of bars: (a) point bar, (b) alternating, (c) tributary, (d) transverse, (e) midchannel. (Illustration by Mark Briggs)

including bar formation, changes in the cross-sectional shape of the channel, and changes in the streambanks (Heede 1980). Bar formations are highlighted here due to their importance in the life cycle of many riparian plant species.

Most bars are longitudinal, in-channel features formed when sediment is deposited along a stream channel due to a drop in local competence (Hedman et al. 1972). Generally, bars are composed of sand or gravel, or a mixture of the two. Bars influence the geometry of flow by changing their size, height, or location in response to flow conditions. Since the form of a channel bar is determined largely by streamflow, the type of bar that occurs in a given situation is an indicator of streamflow and channel conditions (Heede 1980).

Bars are classified as point, alternating, tributary, transverse, and midchannel (fig. 6.2). A point bar occurs when sediment picked up on the concave side of a stream meander, where water velocity is greatest, is replaced by deposition on the convex side as water velocity slows (McBride and Strahan 1984). The greatest accumulation occurs on the convex side just downstream from the position of maximum curvature. The result of this process is meandering or lateral stream movement (McBride and Strahan 1984). Often, a deep pool is formed at the stream's concave bank, and a bar at the convex bank (Rzhanitsyn

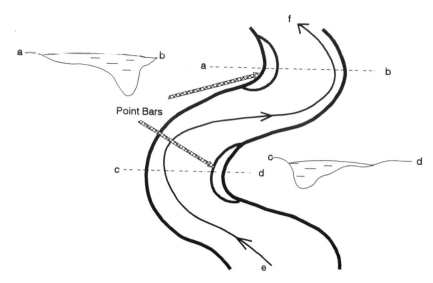

Figure 6.3 Changes in cross-section profile between gentle (a,b) and sharp (c,d) bends. Observe the increase in pool depth with decrease in curvature radius. (Illustration by Mark Briggs; adapted from Heede 1980)

1960). Generally, sharp bends will have deeper pools than moderate bends (fig. 6.3). Heede (1980) noted that although a bar's shape may change with changing flow conditions, the bar itself does not move relative to the bend.

Changes in Channel Longitudinal Profile. The longitudinal profile of a stream channel is essentially the gradient distribution along its length and describes elevation changes of the channel bed over distance (see fig. 2.4). Generally, stream gradient decreases from the headwater reaches to the mouth. However, gradient steepening or lowering can occur along any reach, especially if there are intensive impacts from human activities or natural controls such as rock outcrops or abrupt geologic changes that prevent other types of channel adjustment (Heede 1980).

Bed profile changes can be initiated as a stream cuts its meanders and straightens. The increased concentration of flow can downcut the channel bed, causing the stream channel to incise into valley alluvium (Wolman and Leopold 1957; Burkham 1976; Harvey and Watson 1986), lowering the level of the channel bed. Bed degradation usually precedes profile changes and leads to further channel widening, which occurs when streambanks fail (Schumm et al. 1984; Harvey and Watson 1986) after a critical bank height has been exceeded (Thorne 1981; Little et al. 1982). Bed degradation produces knick-points that migrate upstream, incising the channel beds of increasing numbers of tributaries (Schumm et al. 1984).

Incision into valley alluvium leads to the formation of terraces, abandoned alluvial floodplains no longer affected by annual floods. This changes the topography of the channel from broad and shallow (when excess sediment is stored) to narrow and deep (Wolman and Leopold 1957; Burkham 1976; Schumm et al. 1984; Harvey and Watson 1986). As flow becomes confined to the narrow dimensions of the incised channel, the stream becomes ever more efficient at scouring its bed and banks (Elliott 1979; Van Haveren and Jackson 1986).

Channel incision can significantly affect the amount of water that is available to bottomland vegetation. As channel incision proceeds, there is a lower probability that floods will inundate the entire valley floor (Elliott 1979). Precipitation falling on the watershed spends less time in the fluvial system due to lack of lateral dispersion onto adjacent floodplains (Glinski 1977). A decrease in travel time may also reduce aquifer recharge rates. Degradation of the channel bed also lowers the local groundwater to about the depth of incision in the main channel (Van Haveren and Jackson 1986).

Stream gradient is also influenced by sediment load. The processes by which streams alter their gradient in response to changes in sediment load involve aggradation and degradation during short time periods and changes in sinuosity over longer time periods.

Aggradation occurs when sediment inflow to a reach exceeds the stream's carrying capacity. Sediment drops out and is deposited until a new gradient is established that can carry the incoming sediment (Suryanarayana 1969). The channel will readjust its gradient in a downstream direction as lower reaches adjust to changes upstream. The rate of aggradation decreases with time as gradient changes move downstream; aggradation eventually ceases when all reaches allow equilibrium sediment discharge (Heede 1980).

During degradation, material is picked up from the bed until the carrying capacity of the flow is reached. The elevation of a streambed declines slowly as degradation takes place, and with time the propensity for flow to spill over the banks decreases (Heede 1980).

Of course, the opposite is true when aggradation occurs. Due to decreased flow depth and velocity, the sediment-carrying capacity of floodwaters greatly decreases when they spill over banks. As a result, sediment drops out and is deposited on the banks, and the banks begin to rise. Aggradation may lead to problems if the banks do not rise sufficiently to contain large flood events, making portions of the floodplain more susceptible to the effects of floods that were once contained before the reach started to aggrade.

Strategies for Evaluating
Channel Stability

The reach of the drainageway that passes through the degraded riparian eco-system cannot be evaluated in isolation from the rest of the drainage system. The entire watershed upstream from the degraded riparian site should be examined to identify land use activities (e.g., dams, livestock grazing, con-struction) that affect a stream's flow regime and sediment load. In addition, channel reaches downstream from the site should be examined to identify channel adjustment processes that may migrate upstream and affect the results of the recovery effort (e.g., knickpoints).

Despite the complexity of fluvial adjustments, Elliott (1979), Love (1979), and Schumm et al. (1984) have shown that the evolutionary sequence of events that takes place when a stream adjusts its morphology to accommodate changes in sediment and water discharge can be predicted. These adjustment processes slow with the passage of time, eventually reaching the point where recovery efforts can be effective (Begin et al. 1981). Channel adjustment pro-cesses tend to be rapid where gradients are steep and banks are predominantly of material that can be eroded easily (e.g., sand). The time it takes a particular channel to move through the different stages of adjustment depends primarily on discharge, sediment load, bed and bank stability, tributary influences, and channel gradient (Schumm et al. 1984).

Two evaluation approaches are discussed here: monitoring and instantane-ous. Monitoring involves evaluating channel conditions over an extended pe-riod of time. Instantaneous measurements, by contrast, can be done in a single site visit. Both methods have advantages and disadvantages.

Monitoring channel characteristics over an extended period of time pro-vides the more accurate and realistic picture of the drainageway's geomorphol-ogy. Time constraints are its greatest disadvantage. An effective monitoring program requires years — an option that may not be acceptable to ecosystem managers who need information quickly.

The reverse is true for instantaneous measurements. Instantaneous mea-surements essentially involve translating evidence of form into a prediction of process and change. For example, channel geometry measurements can be used to learn about a drainageway's flow history. Such measurements, how-ever, will not provide as complete a picture as monitoring.

Evaluating Sediment Movement

Evaluating how sediment moves through a channel reach and how sediment storage changes over time can help ecosystem managers identify areas of decreased channel stability. A channel reach with sediment problems (one that is experiencing aggradation or degradation of sediment over a significant time period) can be an indication that the channel is unstable and possibly should be avoided as a revegetation site. The lower reach of Rincon Creek (case study 1; see chapter 2), for example, has experienced significant aggradation over a period of several years, which has produced multiple thalwegs and the propensity to spread out over a wide area.

Using Channel Geometry Measurements to Determine Sediment Storage Changes. Determining the geometry of a channel and how it changes over time can provide information that describes changes in sediment storage. Channel geometry data describe the profile of the channel, which is determined by measuring the vertical (elevation) change with horizontal distance across a cross section of a channel, perpendicular to flow. Channel instability can be detected if repeated surveys over a period of five to ten years of several cross sections along a significant portion of the drainage network reveal significant changes in channel stability.

Ideally, several reaches should be surveyed so that a more comprehensive picture of sediment movement along the drainageway can be obtained. Channel geometry measurements can be taken from one of three geomorphic reference points: the depositional bar, the active-channel level, and the bankfull level (fig. 6.4). The highest level at which channel geometry measurements are commonly made is the bankfull stage (Hedman and Osterkamp 1982). The bankfull stage is the level of the active floodplain at which overbank flooding occurs (Wolman 1955). For many streams in the western United States, however, depositional bar and bankfull references are difficult to evaluate. Depositional bars often do not form in ephemeral and intermittent streams, and measurement of the bankfull reference level requires recognition of a floodplain level, a feature that is typically absent along incised channels (Hedman and Osterkamp 1982).

In contrast, the active-channel reference level is commonly much easier to recognize and, therefore, much easier to measure. The active-channel level is a short-term geomorphic feature subject to change by prevailing discharges (Osterkamp and Hedman 1977). The lower boundary of the active-channel reference level is actively, if not totally, sculpted by the processes of water and

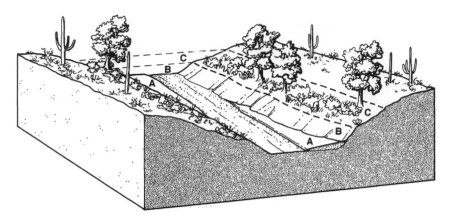

Figure 6.4 Cross section of an idealized stream channel showing three channel geomorphic reference points: (A) depositional bar, (B) active-channel, and (C) bankfull. (Illustration by David Fischer; adapted from Hedman and Osterkamp 1982)

sediment discharge during the normal regime of flow (Wahl 1977). The upper boundary is designated by a break in the relatively steep bank slope of the active channel to a more gently sloping surface beyond the channel edge (Osterkamp and Hedman 1977). This break in slope commonly corresponds with the lower limit of permanent vegetation.

Measuring Channel Geometry. An accurate method for measuring channel geometry is to use a tape to measure horizontal distance, and a level or transit in combination with a graduated staff to measure changes in vertical distance, or channel depth, across the profile of the stream (fig. 6.5). Cross section elevations are referenced to a datum; end points correspond to the bankfull channel and should be marked with monuments. Measurements consist of elevation readings at intervals between the end points. The distance between elevation readings need not be set and can be altered with changes in topography. For example, elevation readings should be more frequent along steep slopes than gentle slopes. A key reference that reviews techniques for gathering data about streams and rivers is Harrelson et al. 1994.

Hedman and Osterkamp (1982) made several recommendations for collecting channel geometry data. First, thoroughly investigate the drainage system to locate at least three cross sections that are representative of the stream channel. Then choose cross sections that are one or more stream widths apart,

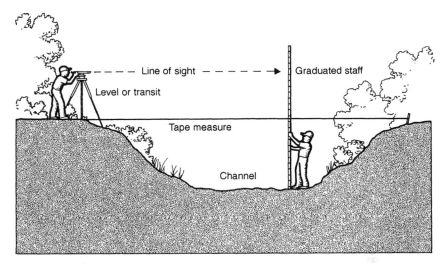

Figure 6.5 Using a level, a graduated staff, and a tape measure is a relatively simple yet accurate way of measuring the dimensions of a channel cross section. (Illustration by David Fischer)

but do not select cross sections immediately up- or downstream from tributaries; such measurements could significantly change the contributing drainage area. Finally, document the procedure by taking a photograph that shows the tape in place.

To determine when changes in channel storage between two time periods were significant, Rhoads and Miller (1990) used a single sample, two-tailed t test ($p=0.20$), with the null hypothesis of no net change in storage within the reach. Significant changes in sediment storage between two different time periods can be an indication of enhanced channel instability, which may indicate the need to consider alternative sites for implementing stream recovery projects.

Using Scour and Fill Chains to Determine Sediment Movement. Scour usually refers to the movement of bed material during high flows, and fill refers to sediment deposition that occurs as floodwaters subside and fill in scoured areas. Recording the amount of scour and fill along various parts of the drainageway over time will provide data describing general aggradation or degradation rates. In addition, scour and fill chains will provide information that describes how sediment is moving during flow.

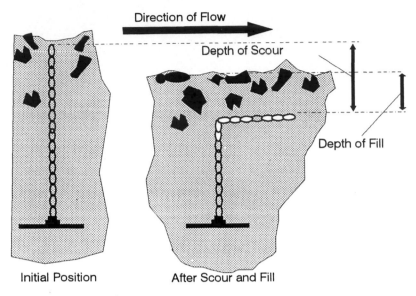

Figure 6.6 Using a chain to measure scour and fill. (Illustration by Mark Briggs)

Scour and fill chains are anchored onto a metal plate and buried in the channel bed (fig. 6.6). Scour chains measure the maximum amount of scour that occurs during a flood (Gordon et al. 1992). As a flood scours away the sediments around the chain, the exposed links fall flat onto the channel bed, forming an abrupt bend. Subsequent filling reburies the chain.

The amount of scour can be determined by measuring the length of the chain above the bend, and the depth of fill can be determined by measuring the amount of sediment above the bend (fig. 6.6). Clearly scour chains can be "lost" quite easily. Make certain that the exact location of the scour chain is described in detail. This is most effectively accomplished by surveying the channel and recording the location of the scour and fill chains in reference to a fixed monument on an upper terrace.

Additional methods for evaluating sediment movement are listed in the following references: Judson 1968 for sediment accretion on archaeological sites, Yair et al. 1978 for measuring the rate of erosion during runoff, Eardley 1966 and Langbein and Schumm 1958 for reservoir sedimentation rate, Walling 1978 for river sediment load monitoring, and Dunne et al. 1978 for tree-root exposure.

Understanding Streamflow

Streamflow data provide an important foundation for understanding a drain-ageway and its associated riparian plant communities. Knowledge of mean discharge is essential for predicting the amount of water that will be available for plant use; flood-frequency discharge combined with channel morphology information can give ecosystem managers a good indication of where vegeta-tion should be placed and which planting sites may be more vulnerable to flood scour than others. Discharge data can be used to estimate sediment yield, analyze the frequency of various flood events, and characterize flow regime in terms of average discharge, or range or variability of discharge (Lewin 1978).

Ecosystem managers are frequently asked to evaluate riparian ecosystems along drainageways where even the most rudimentary hydrologic data are lacking. In such situations, estimates are often made by transferring informa-tion from gaged sites to ungaged sites through regional relations between flow characteristics and the physical and climatic characteristics of the basins. Un-fortunately, flows in arid and semiarid areas are only poorly related to the size of the basin and other basin characteristics (Wahl 1977). Therefore, many standard measurement techniques cannot be used in arid and semiarid parts of the world, where streamflow is infrequent.

Using Channel Geometry Measurements. Relations between stream channel geometry and flow offer a quick and inexpensive alternative method for esti-mating streamflow of ungaged streams. Streamflow characteristics can be ex-trapolated from channel geometry characteristics because the size of an al-luvial channel is indicative of the water conveyed through that channel, while the shape of the channel is for the most part determined by the amount and sizes of sediment transported by the stream (Hedman and Osterkamp 1982). This method is easily applied, requiring only data that describe the channel geometry and the particle size distributions of the material forming the chan-nel perimeter.

Channel geometry data should be collected using recommendations out-lined earlier in this chapter. In addition, particle sizes of the material forming the channel perimeter need to be determined. Three composite samples should be collected from the perimeter of the active channel. The number of samples that make up each composite sample will vary depending on channel size. One composite sample should consist of material collected at equal intervals (every 2 m is recommended) across the channel bed, and the other two composite

Table 6.1 Equations for Determining Mean Annual Runoff for Streams in the Western United States

Flow Frequency	Areas of Similar Regional-Runoff Characteristics	Percentage of Time Having Discharge	Channel-Material Characteristics[a]	Equation[b]	Standard Error of Estimate
Perennial	alpine	>80	silt-clay and armored	$Q_A = 64W_{AC}^{1.88}$	28%
Intermittent	plains north of latitude 39°N	10–80	silt-clay and armored	$Q_A = 40W_{AC}^{1.80}$	50%[c]
			sand	$Q_A = 40W_{AC}^{1.65}$	
	plains south of latitude 39°N	10–80	silt-clay and armored	$Q_A = 20W_{AC}^{1.65}$	50%[c]
			sand	$Q_A = 20W_{AC}^{1.55}$	
Ephemeral	northern and southern plains	6–9	silt-clay and armored	$Q_A = 10W_{AC}^{1.55}$	d
			sand	$Q_A = 10W_{AC}^{1.50}$	d
	intermontane areas	2–5	silt-clay and armored	$Q_A = 4.0W_{AC}^{1.50}$	40%[c]
			sand	$Q_A = 4.0W_{AC}^{1.40}$	40%[c]
	deserts of the Southwest	≤1	silt-clay and armored	$Q_A = .04W_{AC}^{1.75}$	75%[c]
			sand	$Q_A = .04W_{AC}^{1.40}$	75%[c]

[a]Silt-clay channels — bed material $d_{50} < 0.1$ mm or bed material $d_{50} \leq 5.0$ mm and bank silt-clay content $\geq 70\%$.
  Sand channels — bed material $d_{50} = 0.1$–5.0 mm and bank silt-clay content $< 70\%$.
  Armored channels — bed material $d_{50} > 5.0$ mm.
[b]Active-channel width, W_{AC}, in feet; discharge, Q_A, in acre-feet per year.
[c]Approximate standard error of estimate of the basic regression equation.
[d]Standard error of estimate not determined.
Source: Hedman and Osterkamp 1982:13.

samples should consist of samples taken at equal intervals up each bank to the active channel reference point (Osterkamp 1979).

For each of the three composite samples, a particle-size analysis (Guy 1969) is performed to determine three groups of channels: (1) silt-clay channels, those with a median particle size (d_{50}) of bed material less than 0.1 mm or a silt-clay content of 70% or more in bank material and a d_{50} of bed material no greater than 5.0 mm; (2) sand channels, those with d_{50} of the bed material ranging from 0.1 to 5.0 mm and silt-clay contents of the banks of less than 70%; and (3) armored channels, those with d_{50} of the bed material greater than 5.0 mm (Hedman and Osterkamp 1982).

Mean annual discharge can then be calculated by using equations developed by Hedman and Osterkamp (1982). Equations are categorized by flow frequency (perennial, intermittent, and ephemeral) for estimating mean annual runoff and flood discharges with selected recurrence intervals (Table 6.1). The equations take into account the effects of channel material and runoff characteristics.

Active-channel geometry measurements can also be used to describe flood-frequency discharges for streams in the western United States. This type of information can be extremely important to ecosystem managers who are trying to determine the susceptibility to flood scour of projects placed in or near the channel and also to determine general water availability for planted vegetation. Hedman and Osterkamp (1982) developed equations for predicting flood-frequency discharges for alpine streams, streams in the northern plains and intermontane areas east of the Rocky Mountains, streams in the southern plains east of the Rocky Mountains, and streams in the plains and intermontane areas west of the Rocky Mountains (Table 6.2).

Using Crest Stage Gages. Crest stage gages mark the highest elevation of streamflow, allowing ecosystem managers to record peak flows without being present at the site (Harrelson et al. 1994). Crest stage gages are simple devices that consist of a hollow tube (usually metal or PVC) mounted on a strong post that is anchored in the streambed. The tube is open on the bottom and contains a staff that accurately records the high-water line (fig. 6.7). For the crest stage gages installed in Rincon Creek, staffs were covered with Velcro strips and small Styrofoam beads (Styrofoam "peanuts" pulverized in a blender) were placed in the tube. When a flood wave passes, the water rises and the Styrofoam pieces adhere to the Velcro strips. As the flood recedes, some of the Styrofoam pieces will remain on the Velcro, recording the crest stage of the flood. After the high-water mark is recorded, the beads are cleaned off of the Velcro to make ready for the next flow event.

The slope of a flood crest can be developed when several gages are installed along a channel reach. At least three crest stage gages should be installed for each reach where stream discharge information is needed. The slope of the flood peak is obtained by measuring the elevation of the crest flood mark relative to an established reference elevation. Using Manning's equation (see glossary, s.v. "discharge"), slope data can be combined with a roughness coefficient and cross-section data to calculate instantaneous discharge for each of the reaches where the gages were installed. In Rincon Creek, two sets of three gages were installed more than four kilometers apart to obtain information describing the extent to which discharge changes as the creek descends to

Table 6.2 Equations for Determining Flood-Frequency Discharge for Streams in the Western United States

Areas of Similar Climatic Characteristics	Equations	Standard Error of Estimate
Alpine and Pine-Forested	$Q_2 = 1.3W_{AC}^{1.65}$	44%
	$Q_5 = 2.8W_{AC}^{1.60}$	37%
	$Q_{10} = 4.4W_{AC}^{1.55}$	38%
	$Q_{25} = 7.0W_{AC}^{1.50}$	42%
	$Q_{50} = 9.6W_{AC}^{1.45}$	45%
	$Q_{100} = 13W_{AC}^{1.40}$	50%
Northern Plains and Intermontane Areas	$Q_2 = 4.8W_{AC}^{1.60}$	62%
East of Rocky Mountains	$Q_5 = 24W_{AC}^{1.40}$	42%
	$Q_{10} = 46W_{AC}^{1.35}$	40%
	$Q_{25} = 61W_{AC}^{1.30}$	44%
	$Q_{50} = 130W_{AC}^{1.30}$	51%
	$Q_{100} = 160W_{AC}^{1.25}$	58%
Southern Plains East of Rocky Mountains	$Q_2 = 7.8W_{AC}^{1.70}$	66%
	$Q_5 = 39W_{AC}^{1.60}$	57%
	$Q_{10} = 84W_{AC}^{1.55}$	56%
	$Q_{25} = 180W_{AC}^{1.50}$	57%
	$Q_{50} = 270W_{AC}^{1.50}$	59%
	$Q_{100} = 370W_{AC}^{1.50}$	62%
Plains and Intermontane Areas West of	$Q_2 = 1.8W_{AC}^{1.70}$	120%
Rocky Mountains	$Q_5 = 7.0W_{AC}^{1.60}$	73%
	$Q_{10} = 14W_{AC}^{1.50}$	60%
	$Q_{25} = 22W_{AC}^{1.50}$	62%
	$Q_{50} = 44W_{AC}^{1.40}$	71%
	$Q_{100} = 59W_{AC}^{1.40}$	83%

[a] W_{AC} = Active-channel width, in feet; Q_n = discharge, (in cubic feet per second), where n is the recurrence interval in years.
Source: Hedman and Osterkamp 1982:16.

the desert floor. Such information will provide insights as to the extent to which flow is lost by infiltration per unit length of stream.

Evaluating Channel Pattern

Alluvial channels naturally change position across a floodplain with time. Bars and islands shift their position within the channel. Changes in channel position occur due to chute cutoff (rare in arid climates), meander shift, bar movement, and channel diversion. The rate at which a channel changes its position

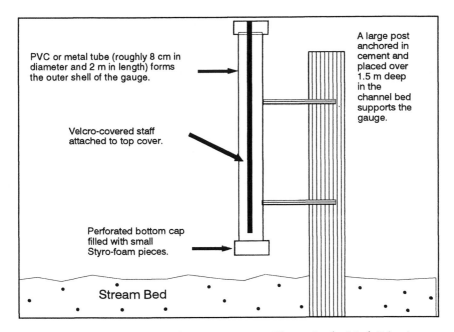

PVC or metal tube (roughly 8 cm in diameter and 2 m in length) forms the outer shell of the gauge.

Velcro-covered staff attached to top cover.

Perforated bottom cap filled with small Styro-foam pieces.

A large post anchored in cement and placed over 1.5 m deep in the channel bed supports the gauge.

Stream Bed

Figure 6.7 Idealized drawing of a crest stage gage. (Illustration by Mark Briggs)

on its floodplain is dependent on many factors, including discharge, sediment supply, alluvial material, streamside vegetation, and the ratio of radius of channel curvature to channel width (Chorley et al. 1984).

Taking into consideration the pattern of a channel can help ecosystem managers compare various sites as to stability. However, evaluating alluvial channel stability needs to be done in the context of the variables that influence the pattern of a channel. Water discharge largely influences channel dimensions, whereas the type and amount of sediment moving through the channel largely influences the shape and pattern of the channel. More precisely, the pattern and shape of an alluvial channel depends largely on the proportion of the total sediment load that is bedload (Schumm 1977).

Channels tend to be narrow and deep (width-to-depth ratio less than 10) when the bedload proportion is small. Depending on its gradient, the channel can be sinuous (low gradient) or straight (high gradient). As the proportion of bedload increases, the width-to-depth ratio of the channel increases and the sinuosity decreases. Multiple thalwegs can form as bedload further increases, and the width-to-depth ratio of the channel will tend to increase and sinuosity will tend to decrease (Chorley et al. 1984).

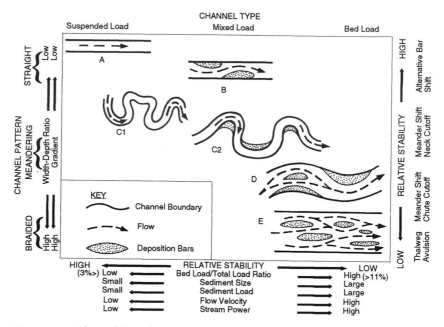

Figure 6.8 Channel classification based on configuration and type of sediment load with associated variables and relative stability indicated. (Illustration courtesy of Schumm [1981, figure 4])

Figure 6.8 depicts six channel configurations and their relative stability. The six channel patterns, of course, do not encompass the overall range of patterns that occur in nature, nor do they show the differences within a pattern (Mollard 1973). They are presented here as a visual review of the nature of channel change, as a way of providing information concerning relative channel stability. (Figure 6.8 pertains solely to unconfined alluvial channels of gentle topographies and is not relevant for assessing high-gradient subalpine streams.) Other geomorphologic characteristics change with changes in channel pattern (see fig. 6.8, lower horizontal axis labels). For example, peak discharge, sediment size, and sediment load all tend to increase from configuration A to configuration E, and relative channel stability decreases from configuration A to configuration E.

Configuration A. The channel is straight with relatively uniform width (fig. 6.8, A). Such channels are usually narrow and deep with low gradients; the proportion of bedload is usually small. Banks tend to be stable due to their high silt-clay content. Depositional bars move through the channel but do not create instability. In general, these channels are not characterized by accelerated bank erosion or rapid channel movements and, at least from the perspective of geomorphology, are sufficiently stable to give planted vegetation a good chance of surviving. Ecosystem managers should, however, be certain that the channel has not been straightened artificially. Artificially straightened channels can move through periods of aggradation and degradation, bank erosion, and scour as they attempt to reestablish their natural position.

Configuration B. The channel is straight but has a sinuous thalweg that carries a small load of coarse sediment that may move through the channel as alternate bars (fig. 6.8, B). At any one location the thalweg will shift with time, leading to situations where deposition on one side of the channel is replaced by scour as alternate bars migrate downstream. This type of configuration is relatively stable. However, revegetation and other recovery strategies should be planned carefully so that they are not destroyed by migrating alternate bars.

Configuration C. This configuration actually represents a continuum of meandering patterns. Two examples are shown (fig. 6.8, C1 and C2). Configuration C1 is a sinuous channel of relatively uniform width that carries a small amount of coarse sediment. Channel banks are relatively stable, but meanders will be cut off. Configuration C2 is also sinuous, but the channel is wider at bends and point bars are large. It is also characterized by relatively rapid meander shifts and neck and chute cutoffs. A shifting meander can alter flow alignment, and bank erosion can become greatly accelerated. The rate at which the meander shifts its position across the floodplain depends on where the river is in the continuum of meandering patterns. Channel reaches with these characteristics are relatively unstable and may not be good candidates for implementing a streamside recovery project.

Configuration D. The channel is relatively wide and shallow (high width-to-depth ratio) with a steep gradient. It is characterized by chute cutoffs, thalweg and meander shift, and bank erosion (fig. 6.8, D). In addition, this type of channel is also characterized by the development of depositional bars that can alter flow patterns, which can change the location of bank erosion. The combination of these problems can make it difficult to find locations suitable for revegetation.

Configuration E. This channel is braided, characterized by bars and thalwegs that shift position frequently (fig. 6.8, E). Gradients are usually steep, and the sediment load and size is large. The channel is unstable and is often characterized by accelerated bank erosion. (Another type of braided stream called an island-braided stream is not depicted in figure 6.8. Island formation, erosion, and shift also occur in these streams but at a much slower rate than in the bar-braided streams.)

Streams in Semiarid Regions and Mountainous Terrain. Figure 6.8 represents streams found more typically in plains regions and valley fill portions of range and basin regions. Many upland streams of semiarid areas and most mountain streams have high gradients, low levels of bedload, and low width-to-depth ratios. Such channels tend to be quite stable because they are highly armored by coarse material.

Developing Recovery Projects along Unstable Alluvial Stream Channels

Addressing the causes of channel instability needs to involve a top-down approach, where upland problems are treated prior to the development of recovery schemes for bottomlands. This can be difficult, however, when instability is the result of land use activities in remote parts of the watershed that may be outside the domain of ecosystem managers or due to land use changes that occurred years ago but whose effects are still being felt. In such cases, recovery may be limited to strategies that address the symptoms of channel instability, and not the causes.

An important factor that needs to be considered is where along the drainageway the recovery will be performed. This is another benefit of evaluating the degraded riparian ecosystem from a watershed perspective. Evaluating the entire drainageway, instead of only the reach that passes through the degraded riparian area, will allow ecosystem managers to prioritize stream reaches so that the ones with the highest priority will receive attention first. Stream reaches that should receive the highest priority are not necessarily the "ugliest," but instead the ones that do not show obvious signs of downcutting and therefore have a greater chance of being effectively treated. The time to implement recovery is before a stream reach shows obvious signs of instability, or after it has reestablished stability. Recovery strategies that are applied locally (e.g., revegetation or the installation of streambank stabilizing structures along the reach that is unstable) may not be effective and may even exacerbate the problem, if the channel is characterized by obvious signs of instability (e.g., headcuts).

Delaying Rehabilitation

Ideally, ecosystem managers should always try to avoid situations where unrealistic deadlines preclude evaluating site characteristics. Postponing recovery projects until after an unstable channel reach has attained a new equilibrium may improve the chances for success (Jackson and Van Haveren 1984; Van Haveren and Jackson 1986; Gore and Bryant 1988). Delaying recovery efforts for even a few years may prevent the development of shortsighted solutions that may either be ineffective or exacerbate the problem (Heede 1986). With more time to evaluate the situation, ecosystem managers will be more likely to develop recovery strategies that address the causes of site decline and work with channel processes, not against them. Some additional time will also allow land managers to better estimate the extent of natural recovery, potentially avoiding the implementation of recovery projects in areas where they are not needed.

Revegetation

As mentioned in chapter 5, planting next to alluvial stream channels can be difficult because there is often a fine line between planting in areas that are too dry and planting in areas that are vulnerable to flood scour. Reichenbacher (1984) described the land surfaces on which riparian communities draw as a continuum. The continuum is least stable near the stream channel where floods are common, and most stable in areas farther removed from the stream channel (e.g., floodplain terraces) where flood disturbances are relatively infrequent. Planting farther away from the channel may provide the newly planted vegetation with the stability required for establishment. However, increased distance from the stream channel is often accompanied by increased depth to groundwater, which can be particularly important if phreatophytes are being used in the revegetation project.

Of course, the ideal planting site is one that offers both protection from flooding and sufficient water. Potential sites include sites in the active channel but just downstream from the concave portion of a channel bend (see chapter 5); downstream from large boulders or other obstacles that can provide

Figure 6.9 Cottonwood poles planted among stabilization jacks for a revegetation project along the Santa Cruz River near Tubac, Arizona. (Photograph by Liz Rosan, Sonoran Institute)

protection from scouring flows; or other areas of the landscape that can provide some protection from flooding.

If sufficient planting materials are available, ecosystem managers may improve their chances of establishing at least some of the vegetation by planting in a variety of different places. For example, planting at various distances from the stream channel may allow the establishment and survival of at least some of the plantings. If weather conditions for the first several years following revegetation are wet, the plantings closest to the active channel may be swept away by flooding, but those planted along higher elevations may survive. If weather conditions are extraordinarily dry, plantings on the higher elevations may desiccate, but plantings in the active channel and lower elevations may still be able to acquire sufficient water.

No matter what the situation, it is always important to plant vegetation in an environment similar to where it is found naturally. Osterkamp and Hupp (1984) found that species distribution is more highly correlated with geomorphic channel features than is sediment size; they inferred that flow frequency and intensity play important roles in determining the distribution of bottomland vegetation. To determine the most suitable planting sites, ecosystem managers must attempt to match planting materials to the geomorphic landscape on which they are naturally found. Most willows (*Salix* spp.), for example, should be planted in dynamic landscapes near the stream channel, settings not suited to species such as catclaw *(Acacia greggii)*, a drought-tolerant species that is commonly found on less dynamic landscapes (e.g., abandoned terraces).

Despite the risks, revegetation offers several advantages over in-stream structures with regard to addressing channel instability and accelerated erosion. In contrast to structures, riparian vegetation can maintain itself in perpetuity via natural propagation and allow streams to function in ways that in-stream structures cannot. Plants maintain the riparian system's resiliency by allowing the system to withstand a variety of environmental conditions (Elmore and Beschta 1987). In addition, revegetation is usually less expensive than installing in-stream structures.

Using revegetation with other recovery strategies should also be considered (fig. 6.9). Goldner (1981) made some observations concerning revegetation along flood control projects. He noted that if flood control projects are going to be revegetated, they need to be constructed with sufficient extra capacity to allow for the growth of planted materials within the channel. If sufficient space is not allowed, excessive growth of vegetation on the channel bottom can lead to logjams and obstructions during high flows, possibly initiating

Figure 6.10 Large tree trunks installed in streambanks were effective in Canyon Creek, Tonto National Forest, Arizona, in promoting alluvial deposition and stream-bank stability and in improving habitat for brown trout. (Photograph by Mark Briggs)

instability up- or downstream. Goldner (1981) also noted that understory vegetation can be included along the lower portions of the channel banks, but trees have to be planted near the top of the banks.

Using Streambank Stabilizing Structures

Streambank stabilization structures include gabions, rip-rap, and check dams (fig. 6.10). These structures are generally designed to slow bank erosion and channel incision. Installation of in-stream structures should be done with utmost caution. By their very nature, in-stream structures alter the natural flow of a river system, initiating changes both up- and downstream from where they are positioned. Dams, for example, slow water flow, allowing sediment to drop out upstream, often producing consequences far removed from the immediate area. Building expensive in-stream structures without addressing impacts associated with the management of riparian ecosystems or uplands may not bring about the intended results (Elmore and Beschta 1987).

For several reasons, the installation of in-stream structures should be one of the last recovery strategies considered. First, relative to many other rehabilitation strategies, installing such structures is often more expensive, requiring large amounts of labor and time. Second, these structures often do not directly address the causes of instability and tend to influence only a limited portion of the channel system. Third, if not installed carefully, these structures can actually perpetuate instability. Placing permanent structures in a channel is an attempt to lock the stream into a fixed location and condition, an unnatural condition that can increase bank erosion and may even lead to the destruction of the structure.

Designing in-stream structures is beyond the scope of this book; the following information may be helpful. Patterson et al. (1981) summarized the effectiveness of various streambank protection techniques currently used by the Soil Conservation Service. Schultze and Wilcox (1985) discussed the results of 29 riparian restoration projects in California's Central Coast area. These projects combined revegetation with four types of structural techniques: bank shaping only, pipe and wire revetment, gabions, and rock rip-rap. The key to success for the majority of these projects was to stabilize the site prior to planting. Sites that were stabilized provided suitable areas for natural revegetation even if artificial planting efforts fail.

Increased salinization is becoming a major environmental problem in many arid and semiarid parts of the world. Whereas salts in humid regions are leached through the soil profile by continuous precipitation, salts in arid regions accumulate through the processes of evaporation and upward capillary movement of moisture from the zone of saturation (Chapman 1975). In his review of soil salinity literature, Hedlund (1984) noted that 222,750 ha of irrigated land in the arid parts of the world are going out of production each year due to salt buildup.

High soil salinity is not common in healthy, lotic riparian ecosystems where annual spring floods remove excess salts. However, the construction of dams along many drainageways has altered natural flow regimes to the point where salt accumulation in soils is contributing to the decline of riverside vegetation communities. In parts of Victoria, Australia, for example, abnormally high levels of salinity are now a common characteristic of many rivers, streams, and wetlands near areas that have been extensively modified by human development (Hart et al. 1990).

In the southwestern United States, agricultural practices within the Colorado River basin have greatly affected the water quality of the Colorado River, which picks up roughly 10 million tons of salt per year as it traverses the seven basin states (Hedlund 1984). Roughly half of the salt comes from natural sources, including mineral springs and geysers, and is continually washed into streams and rivers as they pass through ancient marine deposits where salts accumulate due to lack of leaching and restricted drainage (Chapman 1975). Human inputs from agricultural irrigation (roughly 1 million ha are currently being irrigated by Colorado River water), municipal and industrial use, and reservoir evaporation add the remaining the salts that are picked up by the Colorado River as it moves south to the Mexican border (Jonez 1984).

Unfortunately, the Colorado River is not the only drainage system in the United States with salinity problems. The Rio Grande and Pecos Rivers, as well as the closed river systems in the Great Basin, the Arkansas River, areas of Texas and Oklahoma, and tributaries of the upper Missouri River, also have problems with increased salinity levels (Hedlund 1984).

Soil salinity can reach high levels in riparian areas where groundwater is near the soil surface and where stream waters are high in total dissolved solids (TDS) (Anderson et al. 1984). These characteristics unfortunately fit several rivers in the United States. Salt accumulation in floodplain soils can be especially rapid along drainages whose flood patterns have been artificially altered by impoundment. Along reaches of the lower Colorado River, which has been subjected to the effects of three large dams, buffered spring flows are no longer capable of flushing accumulated salts from many parts of the formerly active floodplain. This greatly increases the likelihood that salts will accumulate to the extent that salinity will negatively affect establishment and growth rates of riparian species.

Effects of Soil Salinity on Plant Growth

Salt accumulation in the soil can produce toxic concentrations of ions that can restrict plant growth (Gauch and Wadleigh 1945). The effects of salinity on plants depend not only on the tolerance of the plant species to salinity, but also on numerous other factors, including climate, amount of soil water, salt composition, soil texture, and stage of development. In general, plants that are highly sensitive to salinity show reduced growth rates in soils that have salinity levels as low as 2 dS m^{-1} (2 decisiemens per meter; see glossary, s.v. "specific electrical conductance"), while plants that are tolerant of salinity show no ill effects in soils with salinity levels as high as 10 dS m^{-1}. Most plants fall between the two extremes, showing slight to moderate sensitivity at 5–10 dS m^{-1}. Plants suffering from salt damage may exhibit leaf scorching (an indication of sodium or chloride toxicity) or leaves that prematurely yellow, wither, or drop. Frequently, however, plants simply appear dull and dark and show no indication of salt damage except stunted growth (Bernstein 1964).

Hale and Orcutt (1987) noted that there are two ways plants become stressed in a saline environment: (1) a reduction in the amount of water that is available to the plant due to the increase in the osmotic potential of the rooting medium; and (2) toxic stress due to the effect of high concentrations of ions. In addition, high concentrations of sodium ions can negatively impact the physical characteristics of a soil, indirectly affecting plant health. Sodic soils readily lose their structure and become impermeable, further reducing the amount of water available for plant use.

Concern over the effects of higher soil salinity levels on agricultural production has focused much of the salinity research on tolerances of agricultural crops to salinity. As a result, salinity tolerances of many nonagricultural plants have not been well documented (Francois, pers. comm., 1993). However, the limited work that has been done on salinity tolerances of nonagricultural species indicates that many riparian species native to the western United States have low salinity tolerances (Anderson 1989; Fenchel et al. 1989; Jackson et al. 1990; Anderson and Taylor, unpub. report, 1992; Pinkney 1992). Indeed, Pinkney (1992) noted in his review of revegetation projects along the lower Colorado River that most native riparian species of this region have low tolerances to salt and that the amount of salt in the soils is therefore a key factor for determining the effectiveness of riparian revegetation. At a riparian revegetation project along the Pecos River in New Mexico, for example, cottonwood and willow poles did not establish, primarily because groundwater and soils were characterized by salinity levels that exceeded 6,000 ppm (roughly 9.2 dS m^{-1}) (Swenson and Mullins 1985).

The Soil Survey

Testing the salinity of a site's soils will help ecosystem managers determine the appropriateness of using revegetation, select appropriate plant materials (e.g., replace salt intolerant species with more salt tolerant species), and decide where they should be planted. A soil survey will provide the data needed to understand the extent to which salt-affected soils are contributing to site deterioration and how salt-affected soils may affect the results of revegetation. However, soil surveys are expensive and probably should not be done until precursory investigations indicate the need for a more detailed survey.

Precursory Investigations

As with groundwater, channel stability, livestock use, and the other factors discussed in this book, an investigation of the riparian site's soil salinity characteristics should begin with a thorough review of existing literature. Libraries, universities, federal agencies (e.g., Soil Conservation Service, U.S. Geological Survey), state organizations (e.g. department of game and fish), national conservation organizations (e.g., the Nature Conservancy), and private consultants are all potential sources of information describing the soil salinity characteristics of a riparian site.

A simple inspection of site conditions by personnel experienced with soil salinity and its impacts on vegetation can also provide some meaningful information. Clues of high soil salinity such as white-crusted soils or vegetation showing symptoms of salinity — or the lack of clues — can help determine the need for a detailed survey.

If the results of these precursory investigations do not rule out soil salinity as a major contributing factor to site deterioration, ecosystem managers need to consider performing soil and water tests. Such tests should begin with a preliminary survey that involves only a few samples. If the results of the preliminary testing reveal potential soil salinity problems, then more thorough testing is advisable before any type of revegetation effort is initiated.

Developing a Soil Sampling Strategy

One of the principal aims of the soil survey is to develop a map that illustrates the soil salinity characteristics of the riparian site. Combining a soil salinity map with information on the salt tolerances of planting materials will allow ecosystem managers to plant appropriate vegetation in soils characterized by salinity levels that permit growth of these plants. The number of soil salinity categories represented on the soil salinity map is determined largely by the tolerances of the plant materials. For example, if only salt-sensitive species are going to be used, the map can be divided into just two parts: areas where the conductivity of the soil's saturation extract is less than 4 dS m^{-1} and areas where the conductivity of the soil's saturation extract is 4 dS m^{-1} or greater. Further salinity categories should be added if species with various salt tolerances are used.

As with most wetland ecosystems, riparian ecosystems exhibit an annual cycle in salinity, with the lowest level of salinity occurring when precipitation is the greatest (Hart et al. 1990). In addition, soil salinity can vary depending on when the sample is collected, as well as on soil depth. Salinity of surface soils, for example, may increase following the wet season even in areas characterized by low-lying groundwater, as evaporation and capillarity carry salts that had previously leached into the subsoil back to the surface (Roundy et al. 1984). In addition to temporal and depth variations, soil salinity can change significantly from one microsite to the next.

The potentially abrupt temporal and spatial changes in the soil salinity of a riparian site mean that an extensive soil survey (involving many soil samples taken at different times of the year) is required before soil salinity characteristics can be accurately understood. Anderson (pers. comm., 1993) performs soil salinity analysis as a routine part of developing a riparian revegetation design. He begins with a preliminary soil sampling and analysis to avoid performing a more intensive soil analysis in areas where salt-affected soils are not a problem or where soil salinity varies little in the riparian site. The methods used in the preliminary sampling phase are generally the same as for full sampling except that fewer samples are taken (Anderson 1989):

> Place a grid system on a map of the revegetation site that divides the riparian site into 0.4 ha plots.
>
> Randomly choose 50% of the plots for sampling.
>
> For each plot, select 8% of the planting points for soil sampling. Anderson (pers. comm., 1992) noted that he generally plants about 250 trees per hectare, which means he samples roughly 20 planting points per hectare, or 5 planting points for each 0.4 ha plot.

Further soil sampling is probably unnecessary if preliminary sampling results describe a site that shows little variation in soil salinity. If the reverse is true, a more intensive sampling scheme should be implemented so that planting in areas of high soil salinity can be avoided. For a more intensive sampling scheme, Anderson (1989) uses the same sampling methods as outlined for the preliminary sampling scheme, except that 10% of the planting points within each 0.4 ha plot are randomly selected instead of 8%.

Aside from the total number of soil samples that need to be taken, several other important factors need to be considered. These factors include the depth from which the soil sample should be taken, the frequency of sampling, and the methods of measuring soil salinity.

Depth of sampling. Depth of sampling refers to the position in the soil profile from which the soil sample is taken. To a large extent, this is determined by the plant species being used and the type of propagule. If large poles are being used, then samples should be taken at the depth where the base of the pole will be placed. If seeds are going to be used, then the upper part of the profile should be considered. Generally, plant species are most vulnerable to the impacts of salinity during the seedling stage. Therefore, ecosystem managers should be primarily concerned with the upper 1.5 m of the soil profile (Francois, pers. comm., 1992). For every sampling point, Anderson (pers. comm.,

1994) takes two samples: one at a depth of 20 cm and another just above the saturated zone of the soil profile.

Frequency of sampling. Because of the dynamic nature of the soil environment, soil samples should be taken at least several times during the growing season. This will provide the most accurate depiction of the variability in soil chemistry. Measuring soil salinity during the time of the year when it is likely to be the highest will allow ecosystem managers to compare maximum soil salinity levels with the maximum salinity tolerances of the plant materials they plan to use in the revegetation project.

Measuring Soil Salinity. Soils are classified into one of four categories based on their salinity content: normal, saline, sodic, and saline-sodic. The criteria that are most commonly used to classify soil salinity are (1) the salinity of the saturation extract as measured by its electrical conductivity (EC) and (2) the sodium absorption ratio (SAR), which is correlated to the amount of sodium ions (Na+) on the cation exchange complex — that is, the exchangeable sodium percentage, or ESP (U.S. Salinity Laboratory 1954). (See glossary s.v. "specific conductance," "sodium absorption ratio," and "exchangeable sodium percentage.") In many parts of the world, SAR measurements have become more common than ESP, because SAR laboratory analysis is easier to perform. As determined for each soil or its saturation extract, ESP is about 1.5 to 1.7 times larger than SAR (U.S. Salinity Laboratory 1954), although the accuracy of this relationship is being questioned in several parts of the world.

Saline soils contain large amounts of soluble salts — commonly sodium, calcium, and magnesium, with chloride, sulfate, and bicarbonate — and are frequently recognized by a white crust on the surface. An SAR of less than 13 indicates a relatively low level of absorbed sodium, while an EC value greater than 4 dS m⁻¹ reflects a high salt concentration.

Sodic soils have a high concentration of sodium on the soil exchange complex but a low concentration of total salts. In sodic soils, SAR values are greater than 13, reflecting a greater amount of exchangeable Na+ than in saline soils. The high percentage of exchangeable Na+ can reduce a soil's hydraulic conductivity by closing conducting pores through in situ mineral swelling (in soils containing significant amounts of expansible minerals) and by particle dispersion and translocation into conducting pores (more common for soils containing smaller amounts of expansible minerals) (McNeal et al. 1966).

Saline soils are commonly sodic (meaning that they have a large amount of soluble salts with a high concentration of sodium on the exchange complex) as well; however, not all saline soils are sodic, nor are all sodic soils saline. Saline-sodic soils have chemical properties of both saline and sodic soils. The exchangeable Na+ is greater than that of a saline soil (SAR13), but the soluble salt concentration is also high (EC4 dS m⁻¹).

One of the easiest methods for determining soil salinity is to extract a sample of the soil solution and determine its composition in the laboratory. Two extraction methods are frequently used. In one method, a sample of soil is collected from the area of interest within the soil profile. The sample is saturated with water in the laboratory, and the soil solution is removed for analysis. The second method is to measure soil salinity in the field directly.

The development of salinity sensors and vacuum soil water samplers have made in situ salinity measurement possible (Oster and Willardson 1971). Yadav et al. (1979) compared several methods for measuring soil salinity under field conditions and concluded that the soil resistivity procedure is probably the best, because it is simple, practical, and rapid. An inexpensive four-electrode probe for monitoring in situ soil salinity has also been developed and has made accurate in situ salinity measurement possible (Rhoades et al. 1976); applications of this method for measuring and monitoring field salinity are described in Rhoades 1976 and Wolt 1994.

Soil Salinity and Revegetation

Revegetating a riparian area characterized by high soil salinity can be difficult because there is often no opportunity to directly address the causes of elevated soil salinity. Generally, increased soil salinity in a floodplain environment is caused by alterations of the natural flow regime (e.g., river impoundment), land use changes (e.g., agricultural drainage), and climatic change. These causes of elevated soil salinity often affect a large area, leaving land managers little opportunity to address them at the local level.

Ecosystem managers can control two factors, however, when planting vegetation in saline riparian soils: the plant materials chosen for the revegetation project and the soil environment. The soil environment to which the seeds, seedlings, poles, or cuttings are exposed can be influenced by the season and by revegetation methodologies. How well the chosen planting materials are adapted to the environmental conditions of the floodplain will determine establishment rates. Of these two factors, selecting plant materials that are adapted to the soil salinity environment is often more effective and less expensive than trying to alter the soil environment.

Choosing Appropriate Plant Species
and Propagules

For revegetation work along the lower Colorado River, Anderson (1989) noted he never plants salt-sensitive vegetation (e.g., Fremont cottonwoods and Goodding willows) in soils where EC is greater than 2.0 dS m^{-1}: four-year-old trees planted in areas of high salinity (greater than 2.0 dS m^{-1}) had 67% less foliage volume than trees planted at the same time in less saline soils. In such situations, ecosystem managers should choose species that can tolerate high soil salinities. Matching salinity tolerances of the plant materials to the soil characteristics of the site can be difficult because of the lack of salt sensitivity data for many riparian species. Nevertheless, information describing the tolerances of some species to salt are described below.

Riparian plants are generally salt sensitive, with adverse effects occurring when salinity levels reach 2,000 mg/l (Hart et al. 1990; Anderson, pers. comm., 1992). Anderson and Ohmart (1986) observed, with other factors being equal, significant $(p<05)$ impact on the vigor of Fremont cottonwoods *(Populus fremontii)* and willows *(Salix* spp.) when soil EC levels exceeded 2.0 dS m^{-1}. They concluded that foliage losses of 30% after two years and 67% after three years should be expected if cottonwoods and willows are planted in soils with EC levels greater than 2.0 dS m^{-1}.

Felker et al. (1981) studied the effects of increased salinity on six species of mesquite *(Prosopis* spp.) and found that their tolerances to soil salinity varied. Velvet mesquite *(P. velutina)* showed signs of decreased growth at the 12,000 mg/l salinity level, while three other mesquite species *(P. articulata, P. pallida,* and *P. tamarugo)* could tolerate 18,000 mg/l NaCl with no reduction in growth.

To make matters even more complicated, tolerance to soil salinity can vary even among members of the same species depending on where the plant is growing. For example, populations of a plant species growing in a freshwater environment can be less salt tolerant than those growing in a more saline environment (Hart et al. 1990). Another factor that should be considered with riparian plants is the combined effects of salinity and waterlogging on plant health. Salinity and waterlogging can act synergistically in affecting plant growth (Van der Moezel et al. 1988), possibly because salinity interferes with the plant's adaptations to waterlogging (Hart et al. 1990).

Table 7.1 summarizes the results of two studies that examined salinity tolerances of plant species associated with southwestern riparian ecosystems. The tolerance ranges that are provided are those where the greatest growth

Table 7.1 Soil Salinity Tolerances of Selected Species

Species	Electrical Conductivity (dS m⁻¹)	Soil Type
Overstory Species		
New Mexican olive	<1.0–2.9[a]	sandy-loamy
Forestiera neomexicana	650–1,885 mg/l[a]	
Fremont cottonwood	<1.0–2.9[a]	sandy-loamy
Populus fremontii	0–1,500 mg/l[b]	sand
Honey mesquite	1,500–36,000 mg/l[b]	sand
Prosopis glandulosa		
Screwbean mesquite	3.0–7.9[a]	clay loam–clay
Prosopis pubescens	1,500–36,000 mg/l[b]	sand
Goodding willow	0–1,500 mg/l[b]	sand
Salix gooddingii		
Black willow	<1.0–3.1[a]	sandy–clay loam
Salix nigra	650–2,015 mg/l[a]	
	3.0–7.9[a]	
Mountain snowberry	1,950–5,135 mg/l[a]	sandy-loamy
Symphoricarpos oreophilus		
Saltcedar	1,500–36,000 mg/l[b]	sand
Tamarix chinensis		
Understory Species		
Pickleweed	1,500–36,000 mg/l[b]	sand
Allenrolfea occidentalis		
Saltbush	8.0–13.9[a]	sandy-loamy
Atriplex canescens	5,200–9,035 mg/l[b]	
Quail bush	0–18,000 mg/l[b]	sand
Atriplex lentiformis		
Squawbush	<1.0–2.9[b]	sandy-loamy
Rhus trilobata		
Roundleaf buffaloberry	<1.0–3.5[b]	loamy–clay
Shepherdia rotundifolia	650–2,275 mg/l[b]	loam
Arrow-weed	0–6,000 mg/l[a]	sand
Pluchea sericea		

[a]*Source:* Anderson and Taylor (n.d.).
[b]*Source:* Jackson et al. 1990.

rates were attained during the seedling stage. Ecosystem managers should look at these tolerance ranges as approximations that indicate optimum conditions. Many of the species listed in table 7.1 are capable of growing outside of these ranges. For example, several studies have demonstrated that *Atriplex* species can establish and grow under TDS concentrations approaching that of seawater (Glenn and O'Leary 1985). In revegetation efforts along the Colorado River, *Atriplex* species established and grew in areas characterized by moderately high salinity, up to 10,000 mg/l (Anderson and Ohmart 1984 in Kerpez and Smith 1987).

Although specific salinity tolerance data are unavailable for many riparian species, it is reasonable to assume that most native riparian species are salt intolerant, with negative effects on growth and establishment occurring when TDS exceed 1,500 mg/l. This assumption is based on two premises. First, healthy riparian ecosystems are not commonly characterized by high soil salinity; and second, the riparian plants that have been studied are not tolerant of high soil salinity. For additional information concerning the soil salinity tolerances of riparian species, ecosystem managers should consult local plant material centers and organizations involved with revegetation.

Reducing Soil Salinity

Some techniques may help to reduce the impact of salt-affected soils on re vegetation. For example, soil salinity can be altered by leaching (through irrigation) and changing the form of the land. Although these strategies were developed specifically for agricultural purposes, they nevertheless may be beneficial for other uses, including riparian revegetation.

Irrigation. Irrigation can be used to leach accumulated salts from the root zone. Leaching is accomplished when sufficient water is applied so that some percolates through the root zone carrying with it a portion of the accumulated salts. Leaching pre-augered holes (see chapter 5) for 48 consecutive hours can prevent salt damage to newly planted saplings (Anderson et al. 1984). Whatever the preplanting water application rate, it is important to remember that over time, salt removal by leaching must equal or exceed the salt additions from the applied water, or salts will reaccumulate (Ayers and Westcot 1989).

Using irrigation to leach salts through the soil profile can be accomplished only if there is adequate drainage. Salts cannot be leached from the root zone in areas with shallow groundwater. If the groundwater is shallow and contains salts, water rising into the root zone by capillarity can become a continual

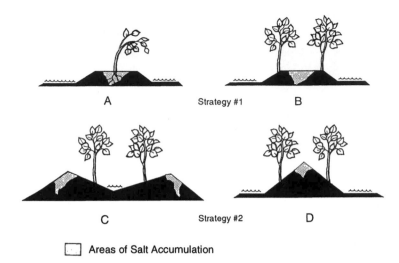

| Areas of Salt Accumulation

Figure 7.1 Planting strategies using raised beds and furrow irrigation: (A,B) salinity tends to increase in the center; (C,D) planting on sloped beds (Illustration by David Fischer; adapted from Bernstein et al. 1955)

source of salts, leading to increased accumulation. In areas of shallow groundwater, the rate of salt accumulation depends on irrigation management, salt concentration and depth of groundwater, soil type, and climatic conditions (Ayers and Westcot 1989).

If there is adequate drainage, the timing and extent of leaching are determined primarily by plant salinity tolerance, soil characteristics, climatic conditions, and quality of irrigation water (Ayers and Westcot 1989). Crops grown in cooler climates or during the cooler time of the year are more tolerant of salinity than are similar crops grown during warm, drier periods. This is primarily because salinity can hamper water uptake by plants, possibly reaching critical levels in warm, dry climates where plant water requirements are significant. In this regard, increasing the frequency of irrigation can help to reduce the likelihood that plants will become stressed from lack of available water.

Land Forming. Land forming can be used to reduce the effects of soil salinity on planting materials. Land forming involves altering the shape of the land; it is usually accomplished with heavy equipment (e.g., a bulldozer) and is therefore not appropriate in all situations. At sites where the top soil layer is characterized by high salinity, Anderson (pers. comm., 1990) recommended using a bulldozer to remove the top 25–30 cm. This strategy prevents irrigation water from transporting salts from the surface down the soil profile to the root zone.

Altering the shape of the seedbed to influence salt accumulation has been used in agricultural practices to improve germination and survival rates of planted species (Ayers and Westcot 1989). This strategy has not been tested in wildland areas and is mentioned only briefly here as a concept that may be worthy of consideration for use in riparian areas where revegetation is being used as a last effort to reestablish native seed sources. However, these strategies are limited to areas where large mechanical equipment and flood irrigation can be used. In addition, all land forming techniques are limited to landscapes that are not frequently influenced by streamflow.

Two strategies for developing raised beds are presented here. Both strategies involve the planting of propagules in raised beds that are irrigated using furrows that lie on both sides of the bed. Water moves from the furrows to the center of the bed. Salts will tend to move with the water, accumulating in the center of the bed. Therefore, a propagule placed in the center of a raised bed that lies between two furrows will be growing in a location where soil salinity will likely be high (fig. 7.1, A).

The first strategy places propagules away from the center of a raised bed (fig. 7.1, B), thereby removing the plantings from the area of greatest salt accumulation (Bernstein et al. 1955). The second strategy involves using sloping beds with the propagules planted just above the water line on the sloping side of the bed (fig. 7.1, C and D) (Bernstein and Fireman 1957). With this strategy, irrigation needs to be continued until the wetting front has moved past the planted vegetation (Ayers and Westcot 1989).

8/ Developing the Recovery Plan

In addition to evaluating the condition of the degraded riparian site, eco-system managers should be familiar with numerous other factors and considerations that can have a tremendous influence on the effectiveness of a recovery effort. Some of these factors are reviewed here.

Developing Project Objectives

The objective of the recovery effort must be stated as clearly as possible from the onset. It is meaningless to evaluate the condition of a degraded riparian area and develop a recovery plan for it without understanding what the end result of the recovery will look like. From his reclamation experiences in Britain, Bradshaw (1988) observed that clearly stated and measurable criteria are necessary not only for choosing among recovery techniques, but also for deciding when a treatment can be declared successful.

Objectives should be developed with the following concepts in mind:

Evaluating site conditions and determining the reasons for decline form the foundation from which realistic objectives can be developed. For example, restoring a riparian area that has experienced significant hydrologic changes to its pre-Anglo-American settlement condition may be next to impossible. In such instances, rehabilitation may be a more suitable objective. At Rincon Creek, for example, revegetating with species that are better adapted to the changed hydrologic conditions instead of using vegetation species that were "originally" at the site is a practical strategy for improving the condition of the site (e.g., planting with mesquite and netleaf hackberry instead of Arizona walnut).

Objectives need to be designed so that they can be translated into measurable criteria. For example, if the objective of the recovery effort is to increase the amount of riparian vegetation on the site, the "amount" desired should be characterized by a descriptor such as numbers of trees, percent cover, density, or volume of vegetation.

Impacts of large-scale recovery projects on water quality, wildlife, and plants need to be considered.

Although postplanting maintenance should be an integral part of any reclamation plan, the long-term objective of the recovery project should be a self-sustaining ecosystem. A self-sustaining ecosystem requires less maintenance and will more likely meet long-term objectives than an ecosystem that is maintained artificially.

Objectives should be developed through an interdisciplinary approach, taking into consideration the opinions of aquatic and riparian biologists, geomorphologists, hydrologists, plant propagationists, soil scientists, watershed managers, engineers, and others.

Objectives need to be based on a realistic understanding of monetary, personnel, and temporal constraints. Although restoration may be what is desired, ecosystem managers may have to settle for a lower level of recovery (e.g., rehabilitation) if resources are insufficient.

Selecting the Best Site

Comparing the pros and cons of several damaged riparian sites in order to select the most suitable one can greatly improve the effectiveness of the recovery effort. In choosing the most suitable site, several important factors need to be considered.

First, prospective sites need to be compared as to their condition. Based on information and procedures documented in the preceding chapters, which of the damaged sites offers conditions most conducive to a successful recovery project? Which of the damaged sites are likely to give the desired results for the least money spent? Which of the sites are most likely to come back on their own, not requiring human intervention? By asking such questions, ecosystem managers can focus on sites where desired results can be achieved most effectively.

Second, each of the potential sites should be considered with the objective of the recovery project in mind. For example, if the goal is to improve wildlife habitat, which site will offer the most benefits to wildlife? Repairing a damaged site that links two healthy natural areas may be more beneficial than repairing a site that is isolated. If the goal is to provide recreational opportunities, which site offers the easiest public access?

Third, potential sites should be compared as to the space they provide for high flow events. Tight canyons characterized by frequent high flow events, for example, may not provide substantial space for enhancing or restoring the riparian zone.

The fourth factor that needs to considered is the past, present, and potential future land uses in and near the sites. A site in an area that is likely to experience considerable changes (e.g., as a result of urbanization) that can degrade the ecological health of the site further may be a less likely candidate than one in a relatively stable area.

Fifth, practical matters should also be considered. How far will project team personnel have to travel? Where will plant materials come from? Access questions, such as road conditions and regulatory constraints, also need to be addressed.

Local, State, and Federal Permit Requirements

Before designing a strategy, spend some time investigating the local, state, and federal permit requirements that may pertain to the site. Ignoring this crucial step may stop a project dead in the water after precious time and money has been spent in developing it. Whether or not the work planned for the site requires a permit depends to a large degree on the site's location, the type of work being planned, what is on the site (e.g., designated critical habitat for endangered species), and who is funding the project. For example, a site may be within the jurisdictional boundary of the U.S. Army Corps of Engineers, potentially requiring a 404 permit. If the site contains cultural resources, a permit from a state agency may be required. In Arizona, for example, the State Preservation Act requires consultation with the state historic preservation officer before projects occurring in areas containing historic cultural resources can proceed. If the site contains habitat critical to an endangered species, a section 7 consultation with the U.S. Fish and Wildlife Service may be required.

Identifying Model Areas

Identifying riparian sites that are similar to the damaged riparian site with regard to elevation, topography, and climate but that are characterized by desired habitat conditions (e.g., hydrologic and vegetation) can provide ecosystem managers with a model that can greatly assist in the development of objectives and methods for in the recovery effort. A plant materials list for revegetating Rincon Creek (case study 1; see chapter 2), for example, will be developed through the examination of a healthy and intact mesquite bosque just downstream from the damaged site. The densities and types of plant species found in the mesquite bosque provide living proof of what is able to survive in the current conditions.

Critical Components of the Recovery Plan

Timetable

A well-planned, realistic timetable can improve the project's chances for success. Ecosystem managers need to allow time for site conditions to be evaluated and the potential for natural recovery to be assessed. If revegetation is going to be used, the timetable needs to take into account the differences between plant species, particularly the time of the year when they should be planted.

Because unforeseen problems and delays are likely to rear their ugly heads, alternative timetables should be developed. In addition, the added expenses that these delays can bring should be incorporated into the project budget — in other words, don't cut the budget so thin that a relatively minor delay could threaten the completion of the entire project (Willard et al. 1990).

Initiating the recovery project during a certain time of the year can greatly increase or decrease a project's effectiveness. Revegetating during the time of the year when precipitation is most likely can greatly improve the chances for establishment, but planting in low-elevation sites during the same time of year can increase the vulnerability of plantings to flood scour. Implementing the project during the time of year when seed fall of desired species occurs can also be of great benefit (Friedman et al. 1995). Finally, minimizing open ground exposure during times of the year when heavy precipitation is likely can help to reduce erosion and other impacts during the construction phase.

Planting Materials

Determining the appropriateness of riparian revegetation and selecting suitable planting materials is determined by a number of different factors, including climate, project objectives, and site conditions (of which drainageway characteristics, water availability, soil conditions, livestock and wildlife use are of particular importance — see relevant chapters in this guidebook).

Ecosystem managers planning to incorporate a revegetation strategy should use only plant materials native to the area. Additionally, ecosystem managers should keep in mind the potential negative effects of bringing off-site species to the revegetation site. For example, collecting cottonwood poles along the

Hassayampa River in central Arizona for a planting project along the Colorado River may produce negative ramifications that are as yet not well understood (e.g., the encroachment of off-site species, to the detriment of on-site species; the introduction of unfamiliar localized diseases or pests). Ecosystem managers should, therefore, always attempt to collect plant materials from on-site or nearby locations. An additional benefit from on-site collection is that plant materials collected from on-site or nearby locations are likely to be better adapted (i.e., more likely to establish and grow) to site conditions than those collected from distant locations. A good reference is Young and Young 1986, which provides information and references for gathering, processing, and germinating seeds of many wildland plants found throughout North America.

The time frame of the project may also determine the species that will be used. If rapid establishment is necessary, planting annuals and biennials may be appropriate, with perennials to be planted at a later date. The availability of planting materials can also limit the quantity and diversity of species used in a riparian revegetation project. Checking on sources of plant materials should therefore be a primary and early consideration when developing recovery strategies.

Documentation

Providing clear and thorough documentation during the planning process is needed to guide project implementation, describe project methods and objectives for the benefit of future team members, help facilitate smooth transitions when personnel changes occur or when the overall plan needs to be revised, and serve as a basis for postproject evaluation.

Budget

Care needs to be taken that all components of the recovery plan have enough funding for implementation. This includes providing for postproject activities (e.g., monitoring, evaluation) that are often slashed to meet budget constraints (Kondolf and Micheli 1995).

Community Involvement

Involving teachers, students, academics, ranchers, landowners, local officials, and other community members in the project from the outset can bring numerous benefits. Pride and the sense of inclusion that community involvement fosters can greatly increase the chances for long-term success. Community members can be involved in a wide range of activities, including revegetation, monitoring, irrigation, installing signs and fences, protecting the site from vandalism, and education outreach activities. Bob Hall (pers. comm., 1992) noted that one of the benefits of riparian revegetation is the potential of involving people of diverse backgrounds directly in the project. His revegetation work along Burro Creek, Arizona, for example, united ranchers and environmentalists (Briggs 1992).

Several organizations across the western United States have been developed specifically to involve communities in identifying and addressing environmental restoration needs. Ecosystem managers are encouraged to contact these and similar organizations to assist in developing community support for their projects. The Adopt-A-Stream Foundation (AASF), based in Everett, Washington, provides funding for community-based habitat restoration and education projects. Using funds from AASF, high school students in Washington have learned how to revegetate streambanks and monitor water quality (Murdoch 1995). The North Portland Youth Conservation Program, a program of the Wetlands Conservancy, teaches young people to identify environmental needs in their community and develop plans for addressing them (Lev 1995).

Demonstration Sites

Developing a demonstration site (essentially a test run of the recovery project that covers a relatively small area) prior to implementing a full-fledged recovery project will allow ecosystem managers to evaluate results and the potential for natural recovery, fine-tune methods, develop more realistic time schedules, promote community participation, add essential project team members, and perform a variety of other activities that can lead to a much more effective project.

Postproject Evaluation and Monitoring

All recovery projects should include postproject monitoring that evaluates the effectiveness of the recovery effort, with the evaluation technique based on the specific project objectives (Kondolf and Micheli 1995). Monitoring the results of the recovery effort allows recovery methods to be adjusted for greater effectiveness, and lessons to be learned from successes and failures can be applied to future efforts. Using appropriate evaluation criteria is crucial in determining whether project objectives have been met (Kondolf and Micheli 1995:4).

Examples of Evaluation Criteria for
Specific Project Objectives

Channel Capacity and Stability. Evaluation criteria include channel cross sections, flood stage surveys, width-to-depth ratio, rates of bank or bed erosion, longitudinal profile, and aerial photography interpretation.

Improve Aquatic Habitat. Evaluation criteria include water depth, water velocities, percentage of overhang cover, shading, pool/riffle composition, stream temperatures, bed material composition, and population assessments for fish, invertebrates, and macrophytes.

Improve Riparian Habitat. Evaluation criteria include percentage of vegetative cover, species densities, size distribution, age-class distribution, the survival of plantings, reproductive vigor, bird and wildlife use, and aerial photography interpretation.

Improve Water Quality. Evaluation criteria include temperature, pH level, dissolved oxygen concentration, conductivity, nitrogen and phosphorus concentrations, the presence of herbicides or pesticides, the turbidity and opacity, the presence of suspended and floating matter, trash loading, and odor.

Recreation and Community Involvement. Evaluation criteria include visual resource improvement based on landscape control point surveys, recreational use surveys, and community participation in management.

Goals of Evaluation Techniques

Evaluation techniques should be designed to generate the most information at the least cost and should be coordinated whenever possible so that data describing several site characteristics can be collected at the same time. For example, study sites along lower Rincon Creek are set up so that channel morphology and streamside vegetation data can be collected using the same transect.

Techniques also need to be reproducible and described in sufficient detail for an outside person to replicate the procedure. Transects or plots need to be permanently established and marked with monuments. Geographical Positioning System (GPS) technology is becoming increasingly commonplace, making it possible to relocate the exact position of transects even when monuments are lost.

Epilogue

The large and growing number of human beings in the southwestern United States ensures that riparian ecosystems in many parts of this region will continue to be extensively damaged. Pressures from development, livestock use, agricultural activities, groundwater pumping, the introduction of flood control structures, and a host of other human-related activities will increasingly affect the condition of these streamside ecosystems. Such a prognosis, combined with the importance of these ecosystems to both wildlife and humans, confirms that developing effective recovery strategies for damaged riparian ecosystems will be increasingly important in the years to come.

Presently, the "science" of repairing ecologically damaged streams and riparian ecosystems is young, and documented experiences from past recovery efforts are scarce. Those experiences that have been documented are therefore inordinately valuable, providing a wealth of information on the types of recovery strategies that are effective (as well as ineffective), the limitations of these strategies, and how their effectiveness can be improved. Drawing lessons from these past experiences is the unifying theme of this book.

One of the critical lessons learned from assessing the effectiveness of past riparian recovery experiences is the importance of evaluating site conditions to understand the causes of ecological decline. Only by understanding the reasons behind the deterioration of the riparian site will ecosystem managers be in a position to develop recovery strategies that directly address the problems.

In this capacity, revegetation is often limited in its ability to improve the condition of damaged riparian ecosystems. Simply replacing lost vegetation with artificially planted species often does not address the causes of ecological decline. Unless revegetation is combined with other recovery methods that address underlying problems, it is prone to failure because the same factors that were responsible for the decline of the riparian area will likely affect the establishment of artificially planted species as well.

Lessons learned from past riparian recovery efforts also show the importance of adopting a broad perspective when attempting to understand riparian ecosystems. Riparian ecosystems are significantly influenced by upstream processes, geological substrates, and the structure and function of upland communities. Thus, riparian systems are susceptible to the impacts of land use over a large region.

One of the major premises for writing this book is that ecosystem managers should evaluate the riparian area from a watershed perspective. At the very least, this necessitates that the evaluation consider the uplands, the tributaries to the drainageway that passes through the site, and the reaches upstream and downstream from the riparian area under study. By stepping away from the bottomland environment and considering the watershed as a whole, ecosystem managers are much more likely to gain from the evaluation process those pieces of information they need to understand the current site condition, the extent to which its function and structure have been changed, and the reasons for the changes that have occurred.

Crucial to this evaluation process is the consideration of factors that have significantly influenced the effectiveness of past riparian recovery efforts. These factors include the extent to which livestock and recreational use are directly affecting the riparian site, the degree to which nonnative plants have become established, the depth to groundwater, the stability of the stream channel that passes through the riparian site, and the salinity of bottomland soils.

Of these factors, evaluating the depth to groundwater and the stability of the stream channel are probably essential components of all riparian recovery efforts that are in low-elevation, alluvial stream channels. For example, depth to groundwater and the extent that groundwater elevation fluctuates during the year will play a large part in determining the types of plant species that can survive. Evaluating the stability of the channel upstream and downstream from the riparian site, as well as the tributaries, will allow ecosystem managers to predict the direction, type, and magnitude of channel adjustment, thereby improving the probability that recovery strategies will work with, not against, natural stream processes. The other factors (livestock grazing, recreational use, nonnative plant invasion, and soil salinity) are certainly important, but their relevance — and therefore the need to evaluate them — will vary from one riparian site to the next.

Another important piece of information that ecosystem managers should ascertain is the potential that the damaged riparian site will recover naturally (i.e., without the aid of human manipulations). In some situations, damaged riparian ecosystems have demonstrated dramatic resiliency, experiencing significant increases in the diversity and density of riparian plants in a short period of time (several years). Understanding the potential for natural recovery will allow ecosystem managers to avoid squandering recovery efforts in riparian areas that are capable of recovering on their own. Just as important, an understanding of natural processes and trends will allow ecosystem managers to work with natural recuperative processes: a highly desirable aim for all riparian recovery efforts.

The evaluation of a damaged riparian site and the development of a recovery plan for it should be a team effort, with input from a multidisciplinary group of individuals that could potentially include specialists in botany, ecology, hydrology, geomorphology, livestock management, and watershed management. In addition, involving the community (e.g., teachers, students, ranchers, landowners) in the riparian recovery effort can be of great benefit. Maintaining an irrigation system, weeding out nonnative species, repairing fences, and preventing vandalism are just some of the advantages that can be accrued when the community is involved.

Clearly, the task of preparing recovery plans for damaged riparian ecosystems is often difficult and challenging. Yet the results of numerous riparian recovery efforts also illustrate the rewards that can be gained when such efforts are effective. Learning the lessons provided by past riparian recovery efforts is one of the keys to enhancing future efforts.

Glossary

alluvium   Generally describes stream-deposited debris.

artificial revegetation (or plantings)   Propagules placed in the ground via human manipulations. This term is predominantly used to emphasize the difference between plant species that establish naturally from those that are planted artificially.

autecology   The study of the ecology of any individual species.

bar   A ridgelike accumulation of sand, gravel, or other alluvial material formed in the channel, along the banks, or at the mouth of a stream where a decrease in velocity induces deposition.

bosque   Spanish for forest; in the Southwest, the term *mesquite bosque* is commonly used to describe bottomland plant communities dominated by *Prosopis* spp.

bottomland   That part of an alluvial valley formed of and underlain by alluvium that has been transported by and deposited by the stream flowing through the valley reach; bottomlands may include the channel bed and banks, bars, and all other alluvial features resulting in a floodplain and one or more terraces.

bottomland ecosystems   The biotic communities associated with rivers, streams, lakes, and other landscape settings where the availability of water is greater than in surrounding uplands. The greater water availability of bottomlands often permits the establishment and growth of many plant species not found on adjacent, drier uplands.

capacity   The total load of sediment that a stream can carry. A stream's capacity increases as its discharge increases.

cation exchange capacity (CEC)   The total of exchangeable cations that a soil can absorb, usually expressed in milliequivalents per 100 g of soil, clay, or organic colloid.

climate   The long-term average weather conditions experienced at a particular place.

competence   The largest particle that a stream can carry. Competence increases as streamflow velocity increases.

cone of depression   The conical-shaped volume of the subsurface around a pumping well that is dewatered owing to the movement of water into the well. When a well is pumped, the elevation of the groundwater drops. The effects of the well on the level of groundwater are greater closer to the well and toward the uppermost level of groundwater. The distance that groundwater is lowered is called drawdown. A drawdown curve shows variation of drawdown with distance from the well.

confined aquifer A groundwater body that is confined by overlying strata of lower permeability. The pressure on this water is greater than atmospheric pressure, so the water levels in wells tapping the aquifer will rise above the bottom of the confining bed.

conservation "Conservation is the state of harmony between men and land" (Leopold 1949:243). For the purposes of this book, this definition also includes actions that will serve to protect land from human impacts and restore areas that have become degraded, while allowing for their sustainable use.

datum A point to which other measurements can be referenced. Often, a datum point is marked in the field with a permanent monument (e.g., a metal or wooden pole placed in the ground to a secure depth).

dioecious Having flowers of one sex, with individual plants being only one sex.

discharge (streamflow) The rate of flow at a given instant expressed as volume per unit of time, generally given in cubic meters per second. Manning's equation is often used to calculate discharge:

$$Q = \frac{1}{n} AR^{2/3} S^{1/2},$$

where Q, n, A, R, and S, respectively, are discharge (m³/s), Manning's roughness coefficient, cross-sectional area of flow (m³), hydraulic radius (m), and slope.

ecosystem The physical and biological features occurring in a given area.

ephemeral stream A stream or reach of a channel that flows only in direct response to precipitation in the immediate locality and whose channel is at all times above the zone of saturation.

exchangeable sodium percentage (ESP) The percentage of the cation exchange capacity of a soil occupied by sodium.

flood Refers to a discharge of greater than bankfull capacity.

floodplain A strip of relatively smooth land bordering a stream, composed of sediment carried by the stream and dropped in the slackwater beyond the influence of the swift current of the channel; the level of the floodplain is generally at about the stage of the mean annual flood, and therefore only one floodplain level can occur in a limited reach of bottomland.

gradient The rate of elevation change per unit length of stream channel, generally expressed as a dimensionless number (m/m).

habitat A specific kind of living space or environment for a particular organism (plant or animal); habitat parameters are determined by a combination of abiotic and biotic factors.

hydric Characterized by an abundance of moisture.

hydrophyte A plant that is adapted to growing in water or very wet conditions. Adaptations include reduced root systems, large floating leaves, the presence of aerenchma in the cortex of the roots and stems, and the development of finely divided submerged leaves.

intermittent stream A stream or reach of a channel that flows only during certain times of the year, such as when it receives water from springs or seeps of the zone of saturation.

knickpoint   Any interruption or break in gradient or slope, especially a point of abrupt change or inflection in the longitudinal profile of a stream.

lotic   Pertaining to life related to water, such as a stream, that is influenced by movement predominantly in one direction; lotic environments, therefore, do not include lakes and oceans.

meander   One of a series of somewhat regular sharp, freely developing, and sinuous curves, bends, turns, or windings in the course of a stream.

mesic   An environment that is neither extremely dry (xeric) nor extremely wet (hydric).

mesophyte   A plant growing in medium moisture conditions (i.e., where moisture availability is between that of xeric and hydric environments).

milliequivalents per liter   Milligrams per liter multiplied by the reciprocals of the combining weights (formula weight of an ion divided by its electrical charge) of the ions that are in solution.

monoecious   Having stamens and pistils in separate flowers on the same individual plant.

perched aquifer   An aquifer containing unconfined groundwater separated from an underlying main body of groundwater by an unsaturated zone. These aquifers usually contain only small quantities of water.

perennial stream   A stream or reach of a channel that flows continuously or nearly so throughout the year and whose upper surface is generally lower than the top of the zone of saturation in the areas adjacent to the stream.

phreatophyte   A plant species that obtains a significant portion of the water that it needs to survive from the zone of saturation or the capillary fringe above the zone of saturation. These species are found in areas characterized by shallow groundwater (e.g., bottomland areas).

pioneer species   A species that tends to be among the first to occupy bare ground; these plants are often intolerant of competition and especially of shading, and may be crowded out as the plant community develops.

pole   A branch or trunk of a woody species that does not have roots or aboveground growth. For some species, branches and trunks can be cut into usable sections for propagation. Fremont cottonwood and Goodding willow poles are commonly used in riparian revegetation projects in the arid Southwest. When placed in suitable locations, poles will develop roots and grow to maturity.

potential evapotranspiration   The maximum moisture that will be transferred from the ground to the atmosphere, provided that the available moisture is sufficient. It is determined predominantly by temperature, relative humidity, and wind speed.

propagule   Any structure, sexual or asexual, that can be used to reproduce a plant species. Examples include poles, cuttings, seedlings, and seeds.

recovery   An effort to improve the condition of an ecosystem that is in less than desirable condition. The objective of this effort can be either restoration, rehabilitation, or replacement (see chapter 1).

recurrence interval   The average time interval, generally expressed in years, between occurrences of a hydrologic event of a specified or greater magnitude.

regime   The stream's seasonal pattern of flow over the year. Streamflow regime can have significant influence on in-stream biota, streamside vegetation characteristics, and average and extreme water temperatures.

rehabilitation   An attempt to create an ecosystem that is similar to (but less than) full restoration (see chapter 1). Rehabilitation attempts to repair damaged ecosystem functions with the primary goal of raising the productivity of the ecosystem for the benefit of the local people. Rehabilitation and restoration are similar in that they both use the ecosystem that existed prior to disturbance as a model, adopting its structure and function. They are also similar in that they attempt to recreate self-sustaining ecosystems that have the ability to repair themselves following natural or human disturbance.

replacement/reallocation   No attempt is made to restore conditions to those that were present originally; instead, the original ecosystem is replaced by a different one (see chapter 1). Bradshaw (1988) noted that this type of reclamation is often the easiest to achieve because it does not require an understanding of the subtle characteristics that made up the original ecosystem. However, in contrast to restoration and rehabilitation, reallocation requires a permanent artificial allocation in the form of energy, water, and fertilizer (Aronson et al. 1993).

restoration   An attempt to create an ecosystem exactly like the one that was present prior to the disturbance (see chapter 1). For several reasons this is often the most difficult to achieve. First, predisturbance conditions are often difficult to describe in sufficient detail to achieve complete restoration. Second, disturbances over many years may have changed the ecosystem to such an extent that complete restoration could be impossible. Third, the original ecosystem may have required several centuries (or much longer) to develop, a temporal scale not appropriate for management.

rhizosphere   The part of the soil that is immediately adjacent (within 1 mm) to the surface of plant roots where the plant most strongly affects soil characteristics.

roughness coefficient   A measure of a channel's irregularity and ability to impede flow. Major determinants of channel roughness are channel and bank vegetation characteristics and channel substrate composition. For a channel of given dimensions and gradient, the velocity of flow decreases as roughness increases. Channel roughness is usually represented by "n" in Manning's formula for determining mean river velocity (see "discharge").

saturation paste   The soil solution that is removed from a saturated soil with vacuum filtration.

sinuosity   The ratio of the length of channel to down-valley distance.

snags   Dead, mature trees that are an important component of forest ecosystems. Snags provide valuable habitat for many species of wildlife.

sodium absorption ratio (SAR)   The proportion of sodium to calcium and magnesium on the cation exchange complex. SAR has a good correlation to the exchangeable sodium percentage (ESP) and is used to estimate the exchangeable sodium percentage of a soil's saturation extract; SAR is more common than ESP because it is much easier to calculate:

$$\text{SAR} = \frac{\text{Na}^+}{\sqrt{\text{Ca}^{++} + \dfrac{\text{Mg}^{++}}{2}}},$$

where the symbols refer to ion concentrations in milliequivalents per liter. A small SAR value (<13) indicates a desirable low sodium content.

soil exchange complex   The charged surfaces of the soil that hold cations for exchange with those in solution. Clays and organic matter are the only soil constituents that typically have a charge. The soil exchange complex is important because the cations held on the exchange complex are available for plants and are not lost with leaching water.

specific electrical conductance   The conductance of a cubic centimeter of solution at a standard temperature (25°C). The use of electrical conductance to measure the salt concentration of solutions is a standard method that has been used for years. Specific electrical conductance is expressed in reciprocal ohms (i.e., ohms^{-1}) and is referred to as *mho* (*ohm* spelled backward). For most measurements, the mho is not appropriate to use because it is too large. Instead, the millimho (1 mho 10^{-3}) and the micromho (1 mho 10^{-6}) are used. However, international (SI) units are being used ever more frequently. Therefore, ecosystem managers should also become familiar with the unit of electroconductivity called the siemen: 0.1 siemen per meter (S m^{-1}) = 1 millisiemen per centimeter = 1 decisiemen per meter (1 dS m^{-1}) = 1 millimho per centimeter (1 mmho cm^{-1}).

stocking rates   The number of livestock per area of rangeland.

stream   A general term for a body of flowing water; in hydrology the term is generally applied to the water flowing in a natural channel as opposed to a canal.

streamflow   The discharge that occurs in (and during floods, adjacent to) a natural channel; the term *streamflow* is more general than *runoff* and can be applied to discharge regardless of whether it is affected by diversion or regulation.

synergism   The condition in which the combined action of two or more agents is greater than the sum of their separate, individual actions.

terrace   A valley-contained aggradational form composed of unconsolidated material; it generally defines a long, narrow, relatively level or gently inclined surface bounded along one edge by a steeper descending slope and along the other by a steeper ascending slope. An alluvial terrace generally reflects an abandoned floodplain surface, is always topographically higher than the floodplain, and is inundated by floods of greater magnitude than the mean annual flood.

thalweg   The line connecting the lowest or deepest points along a streambed or valley.

total dissolved solids (TDS)   The total amount of dissolved mineral matter in water. TDS values are widely used in evaluating water quality and are often determined by measuring electrical conductance (or specific conductance).

watershed   The region or area drained by or contributing to the water of a stream, lake, or other body of water. The terms *drainage area, drainage basin,* and *catchment area* are often used synonymously with *watershed.*

water year   The annual period of October 1 through September 30. The reason for using the water year to publish annual hydrologic data, in lieu of the calendar year, is to avoid splitting the flood season between two consecutive years.

wetted perimeter   The distance along the streambed and banks at a cross section where they contact water.

xeric   Having a low or deficient supply of moisture for plant life.

xerophyte   Any plant growing in dry conditions, able to tolerate periods of drought. Xerophytes have several adaptations that allow them to withstand these extreme conditions, including waxy leaves and leaves reduced to spines, the ability to store water, and short life cycles that can be completed when sufficient water is available. In addition, the term is used to describe plants that can grow in alkaline, acid, and saline soils.

Bibliography

Background Literature, by State

General to Southwestern States

Classification, management, and conservation issues: Brown et al. 1977.
Conservation issues and importance: Johnson 1989.
Ecology and evolution of riparian plant communities: Reichenbacher 1984.
Riparian research expertise directory: Tellman and Jemison 1995.

Arizona

Functions and values to wildlife, a literature review: Ohmart and Zisner 1993.
General bibliography: Simcox and Zube 1985.
Flood dynamics of riparian forests: Stromberg et al. 1991.
Classification of riparian forests and scrubland communities: Szaro 1989.

California

A source book on ecological restoration: Berger 1990.
Field guide to common riparian plants: Faber and Holland 1988.
Ecology, management, and conservation: Warner and Hendrix 1984.

Colorado

Management and conservation issues: Bayha and Schmidt 1983.

New Mexico

Classification of riparian forests and scrubland communities: Szaro 1989.

Oregon

Composition, biomass, and autumn phenology of riparian vegetation of the western Cascade Mountains: Campbell and Franklin 1979.

Wyoming

Riparian community classification: Youngblood et al. 1985.

Selected Federal Archives and Other Agency Sources

Climatic Data Center

National Office, Asheville, N.C., 704-271-4800.
Western Region, Reno, Nev., 702-677-3106.

National Archives

National Archives and Records Center, Washington, D.C., 202-501-5400.

National Park Service

Denver Service Center, 303-969-2100.
Southwest Region, Santa Fe, N.Mex., 505-988-6388.
Rocky Mountain Region, Denver, 303-969-2500.
Western Region, San Francisco, 415-744-3876.

U.S. Bureau of Land Management

Arizona State Office, 602-640-0500.
California State Office, 916-978-4743.
Colorado State Office, 303-239-3701.
Idaho State Office, 208-384-3001.
Nevada State Office, 702-785-6590.
New Mexico State Office, 505-438-7501.
Utah State Office, 801-539-4010.
Wyoming State Office, 307-775-6001.

U.S. Fish and Wildlife Service

Pacific Regional Office, Portland, Oreg. (covers Calif., Idaho, Oreg., and Wash.): Ecological
    Services, 503-231-6159; Refuges and Wildlife, 503-231-6121.
Southwest Regional Office, Albuquerque (covers Ariz., N.Mex., Okla., Tex.): Refuges and
    Wildlife, 612-725-3507.

U.S. Forest Service

Region 2, Rocky Mountain, Lakewood, Colo., 303-275-5001.
Region 3, Southwestern, Albuquerque, 505-842-3260.
Region 4, Intermountain, Ogden, Utah, 801-625-5665.
Region 5, California, San Francisco, 415-705-2884.
Region 6, Pacific Northwest, Portland, Oreg., 503-221-4091.

U.S. Geological Survey

National Center, Reston, Va. (aerial photo., eastern U.S.), 703-648-4000.
Photo. Lib., Denver (aerial photo., Rocky Mtns.), 303-202-4200.
Photo. Lib., Menlo Park, Calif. (aerial photo., western U.S.), 415-329-4309.

Photo. Lib., Rolla, Mo. (aerial photo., central U.S.), 314-341-0851.

Earth Resource Observation System, Sioux Falls, S.Dak., 605-594-6511.

References

Alford, E. 1993. Tonto Rangelands: A Journey of Change. *Rangelands* 15:261–268.

Anderson, B. W. 1989. Research as an Integral Part of Revegetation Projects. In *Proceedings of the California Riparian Systems Conference: Protection, Management, and Restoration for the 1990s (September 22–24, 1988, Davis, Calif.)*, coordinated by D. L. Abell, 413–419. USDA Forest Service General Technical Report PSW-110. Berkeley, Calif.: Pacific Southwest Forest and Range Experiment Station.

Anderson, B. W., J. Disano, D. L. Brooks, and R. D. Ohmart. 1984. Mortality and Growth of Cottonwood on Dredge-Spoil. In *Proceedings of the California Riparian Systems Conference: Protection, Management, and Restoration for the 1990s (September 22–24, 1988, Davis, Calif.)*, coordinated by D. L. Abell, 438–444. USDA Forest Service General Technical Report PSW-110. Berkeley, Calif.: Pacific Southwest Forest and Range Experiment Station.

Anderson, B. W., and S. A. Laymon. 1988. Creating Habitat for the Yellow-Billed Cuckoo *(Coccyzus americana)*. In *Proceedings of the California Riparian Systems Conference: Protection, Management, and Restoration for the 1990s (September 22–24, 1988, Davis, Calif.)*, coordinated by D. L. Abell, 413–419. USDA Forest Service General Technical Report PSW-110. Berkeley, Calif.: Pacific Southwest Forest and Range Experiment Station.

Anderson, B. W., and R. D. Ohmart. 1982. *Revegetation for Wildlife Enhancement along the Lower Colorado River.* Boulder City, Nev.: U.S. Bureau of Reclamation, Lower Colorado Region.

Anderson, B. W., and R. D. Ohmart. 1985. Riparian Revegetation as a Mitigation Process in Stream and River Restoration. In *The Restoration of Rivers and Streams,* edited by J. A. Gore, 41–80. Boston, Mass.: Butterworth Publishers.

Anderson, B. W., and R. D. Ohmart. 1986. *Revegetation at Cecil's Pond.* El Paso, Tex.: Boundary Preservation Project, U.S. Section, International Boundary and Water Commission.

Anderson, B. W., R. D. Ohmart, and J. Disano. 1978. Revegetating the Riparian Floodplain for Wildlife. In *Strategies for Protection and Management of Floodplain Wetlands and Other Riparian Ecosystems (Proceedings of the Symposium, December 11–13, Callaway Gardens, Georgia)*, coordinated by R. R. Johnson and J. F. McCormick, 318–331. USDA Forest Service General Technical Report WO-12. Washington, D.C.: U.S. Department of Agriculture.

Anderson, B. W., and J. Taylor. n.d. Results of Plantings at Bosque del Apache National Wildlife Refuge. Unpublished manuscript, Plant Materials Center, U.S. Soil Conservation Service, Los Lunas, N.Mex.

Anseth, B. 1983. Rancher Fences Creek to Slow Erosion. *Rangelands* 5:204.

Archer, S., and F. E. Smeins. 1991. Ecosystem-Level Processes. In *Grazing Management: An Ecological Perspective,* edited by R. K. Heit Schmidt and J. W. Stuth, 109–139. Portland, Oreg.: Timber Press.

Arizona Riparian Council. 1996. *Arizona Riparian Council Newsletter,* 9(1). Arizona Riparian Council, Center for Environmental Studies, Arizona State University, Tempe.

Arizona Rivers Coalition. 1991. Arizona Rivers: Lifeblood of the Desert — A Citizens Proposal for the Protection of Rivers in Arizona. In-house publication, Arizona Rivers Coalition, Phoenix.

Aronson, J., C. Floret, E. Le Floc'h, C. Ovalle, and R. Pontaneir. 1993. Restoration and Rehabilitation of Degraded Ecosystems in Arid and Semiarid Lands 1: A View from the South. *Restoration Ecology* 1:8–17.

Ayers, R. S., and D. W. Westcot. 1989. *Water Quality for Agriculture.* Irrigation and Drainage Paper 29, rev. 1. Rome: Food and Agriculture Organization of the United Nations.

Bahre, C. J. 1991. *A Legacy of Change: Historic Human Impact on Vegetation in the Arizona Borderlands.* Tucson: University of Arizona Press.

Barrows, C. W. 1993. Tamarisk Control 2: A Success Story. *Restoration and Management Notes* 11:35–38.

Bayha, K. D., and R. A. Schmidt. 1983. Management of Cottonwood-Willow Riparian Associations in Colorado. Unpublished manuscript, Wildlife Society, Colorado chapter, Denver.

Begin, Z. B., D. F. Meyer, and S. A. Schumm. 1981. Development of Longitudinal Profiles of Alluvial Channels in Response to Base-Level Lowering. In *Earth Surface Processes and Landforms,* vol. 6, edited by British Geomorphological Group, 49–68. New York: John Wiley & Sons.

Behnke, R. J., and R. F. Raleigh. 1978. Grazing and the Riparian Zone: Impact and Management Perspectives. In *Strategies for Protection and Management of Floodplain Wetlands and Other Riparian Ecosystems (Proceedings of the Symposium, December 11–13, Callaway Gardens, Georgia),* coordinated by R. R. Johnson and J. F. McCormick, 263–267. USDA Forest Service General Technical Report WO-12. Washington, D.C.: U.S. Department of Agriculture.

Berger, J. J. 1990. *Ecological Restoration in the San Francisco Bay Area: A Descriptive Directory and Sourcebook.* Berkeley, Calif.: Restoring the Earth.

Bernstein, L. 1964. *Salt Tolerance of Plants.* USDA Agricultural Information Bulletin 283. Washington, D.C.: GPO.

Bernstein, L., and M. Fireman. 1957. Laboratory Studies on Salt Distribution in Furrow Irrigated Soil with Special Reference to the Pre-Emergence Period. *Soil Science* 83:249–263.

Bernstein, L., M. Fireman, and R. C. Reeve. 1955. *Control of Salinity in the Imperial Valley, California.* U.S. Department of Agriculture, Agriculture Research Station Report 41(4). Washington, D.C.: GPO.

Bernstein, L., and L. E. Francois. 1973. Leaching Requirement Studies: Sensitivity of Alfalfa to Salinity of Irrigation and Drainage Waters. *Proceedings of the Soil Society of America* 37:931–943.

Beschta, R. L., and W. S. Platts. 1986. Morphological Features of Small Streams: Significance and Function. *Water Resources Bulletin* 22:369–379.

Betancourt, J. C., and R. M. Turner. n.d. *Tucson's Santa Cruz River and the Arroyo Legacy.* Tucson: University of Arizona Press, forthcoming.

Bock, J. H., and C. E. Bock. 1984. Effect of Fires on Woody Vegetation in the Pine-Grassland Ecotone of the Southern Black Hills. *American Midland Naturalist* 112:29–34.

Bock, J. H., and C. E. Bock. 1985. Patterns of Reproduction in Wright's Sycamore. In *Riparian Ecosystems and Their Management: Reconciling Conflicting Uses (First North American Riparian Conference, April 16–18, 1985, Tucson, Arizona),* coordinated by R. R. Johnson, C. D. Ziebell, D. R. Patton, P. F. Ffolliot, and R. H. Hamre, 493–494. USDA Forest Service General Technical Report RM-120. Fort Collins, Colo.: Rocky Mountain Forest and Range Experiment Station.

Braasch, S., and G. W. Tanner. 1989. Riparian Zone Inventory. *Rangelands* 11:103–106.

Bradford, J. M., and R. F. Priest. 1980. Erosional Development of Valley-Bottom Gullies in the Upper Midwestern United States. In *Thresholds in Geomorphology,* edited by D. R. Coates and J. D. Vitak, 75–101. Stroudsburg, Penn.: Dowden and Culver.

Bradshaw, A. D. 1988. Alternative Endpoints in Reclamation. In *Rehabilitating Damaged Ecosystems,* vol. 2, edited by J. Cairns, Jr., 69–85. Boca Raton, Florida: CRC Press.

Brady, W., D. R. Patton, and J. Paxson. 1985. The Development of Southwestern Gallery Forests. In *Riparian Ecosystems and Their Management: Reconciling Conflicting Uses (First North American Riparian Conference, April 16–18, 1985, Tucson, Arizona),* edited by R. R. Johnson, C. D. Ziebell, D. R. Patton, P. F. Ffolliot, and R. H. Hamre, 39–43. USDA Forest Service General Technical Report RM-120. Fort Collins, Colo.: Rocky Mountain Forest and Range Experiment Station.

Brayton, D. S. 1984. The Beaver and the Stream. *Journal of Soil and Water Conservation* 39:108–109.

Briggs, M. 1992. An Evaluation of Riparian Revegetation Efforts in Arizona. Master's thesis, University of Arizona.

Briggs, M. K., B. A. Roundy, and W. W. Shaw. 1994. Trial and Error: Assessing the Effectiveness of Riparian Revegetation in Arizona. *Restoration and Management Notes* 12:160–167.

Brokaw, N.V.L. 1985. Treefalls, Regrowth, and Community Structure in Tropical Forests. In *The Ecology of Natural Disturbance and Patch Dynamics,* edited by S.T.A. Pickett and P. S. White, 53–69. New York: Academic Press.

Brookes, A. 1985. River Channelization: Traditional Engineering Methods, Physical Consequences, and Alternative Practices. *Progress in Physical Geography* 9:44–73.

Brown, D. E., C. H. Lowe, and J. F. Hausler. 1977. Southwestern Riparian Communities: Their Biotic Importance and Management in Arizona. In *Importance, Preservation, and Management of Riparian Habitat: A Symposium (July 9, 1977, Tucson, Arizona),* coordinated by R. R. Johnson and D. A. Jones, 201–211. USDA Forest Service General Technical Report RM-43. Fort Collins, Colo.: Rocky Mountain Forest and Range Experiment Station.

Bryant, L. D. 1982. Response of Livestock to Riparian Zone Exclusion. *Journal of Range Management* 35:780–785.

Buma, P. G., and J. C. Days. 1975. Reservoir Induced Plant Community Changes: A Methodological Explanation. *Journal of Environmental Management* 3:219–250.

Burkham, D. E. 1976. Hydraulic Effects of Changes in Bottomland Vegetation on Three Major Floods, Gila River in Southeastern Arizona. *United States Geological Survey Professional Paper 655-J.* 14 pp. Washington, D.C.: GPO.

Burrows, W. J., and D. J. Carr. 1969. Effects of Flooding the Root System of Sunflower Plants on the Cytokinin Content in the Xylem Sap. *Physiologia Plantarum* 22:1105–1112.

Busby, F. E. 1979. Riparian and Stream Ecosystems, Livestock Grazing, and Multiple-Use Management. In *Proceedings of the Forum on Grazing and Riparian Stream Ecosystems (November 3–4, 1978, Denver, Colo.),* edited by O. B. Cope, 6–12. Denver: Trout Unlimited.

Busch, S. E., N. L. Ingraham, and S. S. Smith. 1992. Water Uptake in Woody Riparian Phreatophytes of the Southwestern U.S.: A Stable Isotope Study. *Ecological Applications* 2:450–459.

Campbell, A. G., and J. F. Franklin. 1979. *Riparian Vegetation in Oregon's Western Cascade Mountains: Composition, Biomass, and Autumn Phenology.* Coniferous Forest Biome,

Ecosystem Analysis Studies, U.S./International Biological Program Bulletin no. 14. Corvallis, Oreg.: Forest Service Laboratory.

Campbell, C. J., and W. Green. 1968. Perpetual Succession of Stream-Channel Vegetation in a Semiarid Region. *Journal of the Arizona Academy of Science* 5:86–97.

Carothers, S. W., G. S. Mills, and R. R. Johnson. 1990. The Creation and Restoration of Riparian Habitat in Southwestern Arid and Semi-Arid Regions. In *Wetland Creation and Restoration: The Status of the Science,* vol. 1, *Regional Reviews,* edited by J. A. Kusler and M. E. Kentula, 359–376. Covelo, Calif.: Island Press.

Case, R. L. 1995. Structure, Biomass, and Recovery of Riparian Ecosystems of Northeast Oregon. Master's thesis, Oregon State University.

Chaney, E., W. Elmore, and W. S. Platts. 1990. *Livestock Grazing on Western Riparian Areas.* Eagle, Idaho: Northwest Resource Information Center (prepared for the U.S. Environmental Protection Agency).

Chaney, E., W. Elmore, and W. S. Platts. 1993. *Managing Change — Livestock Grazing on Western Riparian Areas.* Eagle, Idaho: Northwest Resource Information Center (prepared for the U.S. Environmental Protection Agency).

Chapman, V. J. 1975. The Salinity Problem in General, Its Importance, and Distribution with Special Reference to Natural Halophytes. In *Plants in Saline Environments,* edited by A. Poljakoff-Mayber and J. Gale, 7–24. New York: Springer-Verlag.

Chorley, R. J., S. A. Schumm, and D. E. Sugden. 1984. *Geomorphology.* New York: Methuen.

Clary, W. P. 1995. Vegetation and Soil Responses to Grazing Simulation on Riparian Meadows. *Journal of Range Management* 48:18–25.

Cole, D. N. 1993. *Trampling Effects on Mountain Vegetation in Washington, Colorado, New Hampshire, and North Carolina.* USDA Forest Service Research Paper INT-464. Ogden, Utah: Intermountain Research Station.

Cole, D. N. 1995. Disturbance of Natural Vegetation by Camping: Experimental Applications of Low-Level Stress. *Environmental Management* 19(3):405–416.

Craig, W. S. 1977. Reducing Impacts from River Recreation Users. In *Proceedings from the Symposium on River Recreation Management and Research,* 155–162. USDA Forest Service General Technical Report NC-28. St. Paul, Minn.: North Central Forest Experimental Station.

Curll, M. L., and R. J. Wilkins. 1982. Effects of Treading and the Return of Excreta on a Perennial Ryegrass–White Clover Sward Defoliated by Continuously Grazing Sheep. In *Proceedings of the 14th International Grasslands Congress,* edited by J. A. Smith and V. W. Hays, 456–458. Boulder, Colo.: Westview Press.

Curll, M. L., and R. J. Wilkins. 1983. The Comparative Effects of Defoliation, Treading, and Excreta on a *Lolium perenne–Trifolium repens* Pasture Grazed by Sheep. *Journal of Agricultural Science* [Cambridge] 100:451–460.

Dagget, D. 1994. Environmentalist of the Year. *Range Magazine* 2(2): 20–21, 37.

Dalton, M. G., B. E. Huntsman, and K. Bradbury. 1991. Acquisition and Interpretation of Water-Level Data. In *Practical Handbook of Ground-Water Monitoring,* edited by D. M. Nielson, 367–396. Chelsea, Mich.: Lewis Publishers.

Davis, T. 1995. The Southwest's Last Real River: Will It Flow On? *High Country News* 27(11):1, 10–11.

Dotzenko, A. D., N. T. Papamichos, and D. S. Romine. 1967. Effect of Recreational Use on Soil and Moisture Conditions in Rocky Mountain National Park. *Journal of Soil and Water Conservation* 22:196–197.

Duff, D. 1979. Riparian Habitat Recovery on Big Creek, Rich County, Utah: A Summary of Eight Years of Study. In *Proceedings of the Forum on Grazing and Riparian Stream Ecosystems (November 3–4, 1978, Denver, Colo.)*, edited by O. B. Cope, 41–46. Denver: Trout Unlimited.

Dunne, T., W. E. Dietrich, and M. J. Bruneggo. 1978. Recent and Past Erosion Rates in Semi-Arid Kenya. *Zeitschrift für Geomorphologie* Supplement NF 29:130–40.

Eardley, A. J. 1966. Rates of Denudation in the High Plateaus of Southwestern Utah. *Bulletin of the Geological Society of America* 77:777–780.

Elliott, J. G. 1979. Evolution of Large Arroyos: The Rio Puerco of New Mexico. Master's thesis, Colorado State University.

Elmore, W. 1989. Riparian Management: Oregon Recipes. Paper presented at the Fourth Wild Trout Conference, September 18–19, 1989, Yellowstone National Park, Mammoth, Wyoming.

Elmore, W., and R. L. Beschta. 1987. Riparian Areas: Perceptions in Management. *Rangelands* 9:260–265.

Ericsson, K. A., and D. J. Schimpf. 1986. Woody Riparian Vegetation of a Lake Superior Tributary. *Canadian Journal of Botany* 64:769–773.

Everitt, B. L. 1968. Use of the Cottonwood in an Investigation of the Recent History of a Flood Plain. *Science* 266:417–439.

Faber, P. M., and R. F. Holland. 1988. *Common Riparian Plants of California: A Field Guide for the Layman*. Mill Valley, Calif.: Pickleweed Press.

Farmer, R. E., Jr., and F. T. Bonner. 1967. Germination and Initial Growth of Eastern Cottonwood as Influenced by Moisture Stress, Temperature, and Storage. *Botanical Gazette* 128:211–215.

Felker, P., P. R. Clark, A. E. Laag, and P. F. Pratt. 1981. Salinity Tolerance of the Tree Legumes: Mesquite *(Prosopis glandulosa* var. *torreyana, P. velutina,* and *P. articulata)*, Algarrobo *(P. chilensis)*, Kiawe *(P. pallida)*, and Tamarugo *(P. tamarugo)* Grown in Sand Culture on Nitrogen-Free Media. *Plant and Soil* 61:311–317.

Fenchel, G., W. Oaks, and E. Swenson. 1989. *Selecting Desirable Woody Vegetation for Environmental Mitigation and Controlling Wind Erosion and Undesirable Plants in the Rio Grande and Pecos River Valleys of New Mexico*. Five-year interim report. Los Lunas, N.Mex.: Plant Materials Center, U.S. Soil Conservation Service.

Fenner, P., W. W. Brady, and D. R. Patton. 1984. Observations on Seeds and Seedlings of Fremont Cottonwood. *Desert Plants* 6:55–58.

Fenner, P., W. W. Brady, and D. R. Patton. 1985. Effects of Regulated Water Flows on Regeneration of Fremont Cottonwood. *Journal of Range Management* 38:135–138.

Finlayson, C. M., I. D. Cowie, and B. J. Bailey. 1990. Sediment Seedbanks in Grassland on the Magela Creek Floodplain, Northern Australia. *Aquatic Botany* 38:177–193.

Fowells, H. A. 1965. *Silvics of Forest Trees of the United States*. USDA Forest Service Handbook no. 271. Washington, D.C.: U.S. Department of Agriculture.

Friedman, J. M., M. L. Scott, and W. M. Lewis, Jr. 1995. Restoration of Riparian Forest Using Irrigation, Artificial Disturbance, and Natural Seedfall. *Environmental Management* 19:547–557.

Frissell, S. S., Jr., and D. P. Duncan. 1965. Campsite Preference and Deterioration in the Quetico Superior, Canoe County. *Journal of Forestry* 63:256–260.

Gauch, H. G., and C. H. Wadleigh. 1945. The Influence of High Concentrations of Sodium, Calcium, Chloride, and Sulfate on Ionic Absorption by Bean Plants. *Soil Science* 59:139–153.

Gebhardt, K. A., C. Bohn, S. Jensen, and W. S. Platts. 1989. Use of Hydrology in Riparian Classification. In *Practical Approaches to Riparian Resource Management: An Educational Workshop (May 8–11, 1989, Billings, Montana)* edited by R. E. Gresswell, B. A. Barton, and J. L. Kershner, 53–59. Billings, Mont.: U.S. Bureau of Land Management.

Gecy, J. L., and M. V. Wilson. 1990. Initial Establishment of Riparian Vegetation after Disturbance by Debris Flows in Oregon. *American Midland Naturalist* 123:282–291.

Gifford, G. F., and R. H. Hawkins. 1978. Hydrologic Impact of Grazing on Infiltration: A Critical Review. *Water Resources Research* 14:305–313.

Gilbert, G. K. 1880. *Report on the Geology of the Henry Mountains.* 2nd ed. U.S. Geographical and Geological Survey of the Rocky Mountain Region. Washington, D.C.: GPO.

Gill, D. 1973. Modification of Northern Alluvial Habitats by River Development. *Canadian Geography* 17:138–153.

Gillen, R. L., W. C. Krueger, and R. F. Miller. 1985. Cattle Use of Riparian Meadows in the Blue Mountains of Northeastern Oregon. *Journal of Range Management* 38:205–209.

Glenn, E. P., and J. W. O'Leary. 1985. Productivity and Irrigation Requirements of Halophytes Grown with Seawater in the Sonoran Desert. *Journal of Arid Environments* 9:81–91.

Glinski, R. L. 1977. Regeneration and Distribution of Sycamore and Cottonwood Trees along Sonoita Creek, Santa Cruz County, Arizona. In *Importance, Preservation, and Management of Riparian Habitat: A Symposium (July 9, 1977, Tucson, Arizona),* coordinated by R. R. Johnson and D. A. Jones, 116–123. USDA Forest Service General Technical Report RM-43. Fort Collins, Colo.: Rocky Mountain Forest and Range Experiment Station.

Goldner, B. H. 1981. Riparian Restoration Efforts Associated with Structurally Modified Flood Control Channels. In *California Riparian Systems: Ecology, Conservation, and Productive Management,* edited by R. E. Warner and K. M. Hendrix, 445–451. Berkeley and Los Angeles: University of California Press.

Gordon, N. D., T. A. McMahon, and B. L. Finlayson. 1992. *Stream Hydrology: An Introduction for Ecologists.* West Sussex, U.K.: John Wiley & Sons.

Gore, J. A., and F. L. Bryant. 1988. River and Stream Restoration. In *Rehabilitating Damaged Ecosystems,* vol. 1, edited by J. Cairns, Jr., 24–36. Boca Raton, Fla.: CRC Press.

Green, D. M., and J. B. Kauffman. 1995. Succession and Livestock Grazing in a Northeastern Oregon Riparian Ecosystem. *Journal of Range Management* 48:307–313.

Griggs, G. B. 1984. Flood Control and Riparian System Destruction: Lower San Lorenzo River, Santa Cruz County, California. In *California Riparian Systems: Ecology, Conservation, and Productive Management,* edited by R. E. Warner and K. M. Hendrix, 142–150. Berkeley and Los Angeles: University of California Press.

Guy, H. P. 1969. *Laboratory Theory and Methods for Sediment Analysis.* U.S. Geological Survey Techniques of Water Resources Investigations, book 5, C-1. Reston, Va.

Hack, J. T. 1960. Interpretation of Erosional Topography in Humid Temperate Regions. *American Journal of Science,* Bradley Volume 258A:80–97.

Hagen, R. M., and E. B. Roberts. 1972. Ecological Impacts of Water Projects in California. *Proceedings of the American Society of Civil Engineers* 98:25–48.

Hale, G. H., and D. M. Orcutt. 1987. *The Physiology of Plants under Stress.* New York: John Wiley & Sons.

Hall, F. R. 1968. Base-Flow Recession: A Review. *Water Resources Research* 4:973–983.

Hall, R. S., and A. R. Bammann. 1987. Riparian Restoration Techniques on Arizona's Public Lands. Paper presented at the Riparian Revegetation Symposium, April 15–18, 1987, San Diego, California.

Hammitt, W. E., and D. N. Cole. 1987. *Wildland Recreation: Ecology and Management.* New York: John Wiley & Sons.

Hardin, E. D. 1984. Variation in Seed Weight, Number per Capsule, and Germination in *Populus deltoides* Bartr. Trees in Southeastern Ohio. *American Midland Naturalist* 112: 29–34.

Harrelson, C. C., C. L. Rawlins, and J. P. Potyondy. 1994. *Stream Channel Reference Sites: An Illustrated Guide to Field Techniques.* USDA Forest Service General Technical Report RM-245. Fort Collins, Colo.: Rocky Mountain Forest and Range Experiment Station.

Hart, T. B., P. Baily, R. Edwards, K. Hortle, K. James, A. McMahon, C. Meredith, and K. Swadling. 1990. Effects of Salinity on River, Stream, and Wetland Ecosystems in Victoria, Australia. *Water Research* 24:1103–1117.

Harvey, M. D., and C. D. Watson. 1986. Fluvial Processes and Morphological Thresholds in Incised Channel Restoration. *Water Resources Bulletin* 2:359–368.

Heady, H. F. 1975. *Rangeland Management.* New York: McGraw-Hill.

Hedlund, J. D. 1984. USDA Planning Process for Colorado River Basin Salinity Control. In *Salinity in Watercourses and Reservoirs: Proceedings of the 1983 International Symposium on State-of-the-Art Control of Salinity (July 13–15, 1983, Salt Lake City, Utah),* edited by R. H. French, 63–77. Boston, Mass.: Butterworth Publishers.

Hedman, E. R., D. O. Moore, and R. K. Livingston. 1972. *Selected Stream-Flow Characteristics as Related to Channel Geometry of Perennial Streams in Colorado.* U.S. Geological Survey Open-File Report. Washington, D.C.: GPO.

Hedman, E. R., and W. R. Osterkamp. 1982. *Streamflow Characteristics Related to Channel Geometry of Streams in Western United States.* U.S. Geological Survey Water-Supply Paper 2193. Washington, D.C.: GPO.

Heede, B. H. 1980. *Stream Dynamics: An Overview for Land Managers.* USDA Forest Service General Technical Report RM-72. Fort Collins, Colo.: Rocky Mountain Forest and Range Experiment Station.

Heede, B. H. 1981. Rehabilitation of Disturbed Watersheds through Vegetation Treatment and Physical Structures. In *Interior West Watershed Management, Proceedings of a Symposium (April 8–10, Spokane, Wash.),* edited by D. M. Baumgartner, 257–260. Pullman: Washington State University.

Heede, B. H. 1986. Designing for Dynamic Equilibrium in Streams. *Water Resources Bulletin* 22:351–357.

Hendrickson, D. A., and W. L. Minckley. 1985. Cienegas: Vanishing Climax Communities of the American Southwest. *Desert Plants* 6:130–175.

Hennessy, J. T., R. P. Gibbons, J. M. Tromble, and M. Cardenas. 1985. Mesquite *(Prosopis glandulosa Torr.)* Dunes and Interdunes in Southern New Mexico: A Study of Soil Properties and Soil Water Relations. *Journal of Arid Environments* 9:27–38.

Hopkins, R. M., and W. H. Patrick, Jr. 1969. Combined Effect of Oxygen Content and Soil Compaction on Root Penetration. *Soil Science* 108:408–413.

Horton, R. E. 1945. Erosional Development of Streams and Their Drainage Basins: Hydrophysical Approach to Quantitative Morphology. *Bulletin of the Geological Society of America* 56:275–370.

Howard, S. W., A. E. Dirar, J. O. Evan, and F. D. Provenza. 1983. The Use of Herbicides and/or Fire to Control Saltcedar *(Tamarix).* *Proceedings of the Western Society of Weed Science* 36:65–72.

Howe, W. H., and F. L. Knopf. 1991. On the Imminent Decline of Rio Grande Cottonwoods in Central New Mexico. *Southwestern Naturalist* 36:218–224.

Hubbard, J. P. 1977. Importance of Riparian Ecosystems: Biotic Considerations. In *Importance, Preservation, and Management of Riparian Habitat: A Symposium (July 9, 1977, Tucson, Arizona)*, coordinated by R. R. Johnson and D. A. Jones, 14–18. USDA Forest Service General Technical Report RM-43. Fort Collins, Colo.: Rocky Mountain Forest and Range Experiment Station.

Hupp, C. R. 1982. Stream-Grade Variation and Riparian Forest Ecology along Passage Creek, Virginia. *Bulletin of the Torrey Botanical Club* 109:488–499.

Hupp, C. R., and W. R. Osterkamp. 1985. Bottomland Vegetation Distribution along Passage Creek, Virginia, in Relation to Fluvial Landforms. *Ecology* 66:670–681.

Jackson, J., J. T. Ball, and M. R. Rose. 1990. *Assessment of the Salinity Tolerance of Eight Sonoran Desert Riparian Trees and Shrubs*. Final report. Reno: Desert Research Institute, University of Nevada System, Biological Sciences Center.

Jackson, W. L., and B. P. Van Haveren. 1984. Design for Stable Channel in Coarse Alluvium for Riparian Zone Restoration. *Water Resources Bulletin* 20:695–703.

Johnson, A. S. 1989. The Thin Green Line: Riparian Corridors and Endangered Species in Arizona and New Mexico. In *Preserving Communities and Corridors*, edited by G. Mackintosh, J. Fitzgerald, and D. Kloepfer, 35–46. Washington, D.C.: Defenders of Wildlife.

Johnson, R. R., and S. W. Carothers. 1982. *Riparian Habitat and Recreation: Interrelationships and Impacts in the Southwest and Rocky Mountain Region*. Eisenhower Consortium for Western Environmental Forestry Research, Bulletin no. 12. Fort Collins, Colo.

Johnson, R. R., and D. A. Jones, editors. 1977. *Importance, Preservation, and Management of Riparian Habitat: A Symposium (July 9, 1977, Tucson, Arizona)*. USDA Forest Service General Technical Report RM-43. Fort Collins, Colo.: Rocky Mountain Forest and Range Experiment Station.

Johnson, R. R., and J. F. McCormick, technical coordinators. 1978. *Strategies for Protection and Management of Floodplain Wetlands and Other Riparian Ecosystems (Proceedings of the Symposium, December 11–13, Callaway Gardens, Georgia)*. USDA Forest Service General Technical Report WO-12. Washington, D.C.: U.S. Department of Agriculture.

Jonez, A. R. 1984. Controlling Salinity in the Colorado River Basin. In *Salinity in Watercourses and Reservoirs: Proceedings of the 1983 International Symposium on State-of-the-Art Control of Salinity (July 13–15, 1983, Salt Lake City, Utah)*, edited by R. H. French, 337–347. Boston, Mass.: Butterworth Publishers.

Judd, B. I., J. M. Laughlin, H. R. Guenther, and R. Handegarde. 1971. The Lethal Decline of Mesquite on the Casa Grande National Monument. *Great Basin Naturalist* 31:153–159.

Judson, S. 1968. Erosion Rates near Rome, Italy. *Science* 160:1444–1445.

Kauffman, J. B., R. L. Case, D. Lytjen, N. Otting, and D. L. Cummings. 1995. Ecological Approaches to Riparian Restoration in Northeast Oregon. *Restoration and Management Notes* 13:12–19.

Kauffman, J. B., and W. C. Krueger. 1984. Livestock Impacts on Riparian Ecosystems and Streamside Management Implications: A Review. *Journal of Range Management* 37:430–437.

Kearney, T. H., and R. H. Peebles. 1969. *Arizona Flora*. Berkeley: University of California Press.

Keller, E. A., and A. Brookes. 1984. Consideration of Meandering in Channelization Projects: Selected Observations and Judgements. In *River Meandering: Proceedings of the 1983 Rivers Conference*, edited by C. M. Elliott, 384–398. New York: American Society of Civil Engineers.

Keller, C., L. Anderson, and P. Tappel. 1979. Fish Habitat Changes in Summit Creek, Idaho, after Fencing. In *Proceedings of the Forum on Grazing and Riparian Stream Ecosystems (November 3–4, 1978, Denver, Colo.)*, edited by O. B. Cope, 46–52. Denver: Trout Unlimited.

Kerpez, T. A., and N. S. Smith. 1987. *Saltcedar Control for Wildlife Habitat Improvement in the Southwestern United States.* U.S. Fish and Wildlife Resource Publication no. 169. Washington, D.C.

Kerr, J. A. 1973. Physical Consequences of Human Interference with Rivers. In *Proceedings of the 9th Canadian Hydrology Symposium, Fluvial Processes and Sedimentation,* Research Council of Canada. Ottawa: Government Printing.

Kondolf, G. M., and R. R. Curry. 1984. The Role of Riparian Vegetation in Channel Bank Stability: Carmel River, California. In *California Riparian Systems: Ecology, Conservation, and Productive Management,* edited by R. E. Warner and K. M. Hendrix, 124–134. Berkeley and Los Angeles: University of California Press.

Kondolf, G. M., and E. R. Micheli. 1995. Evaluating Stream Restoration Projects. *Environmental Management* 19:1–15.

Langbein, W. B., and S. A. Schumm. 1958. Yield of Sediment in Relation to Mean Annual Precipitation. *Transactions of the American Geophysical Union* 39:1076–1084.

Leopold, A. 1949. *A Sand County Almanac.* New York: Oxford University Press.

Leopold, L. B. 1951. Rainfall Frequencies: An Aspect of Climatic Variation. *Transactions of the American Geophysical Union* 32:342–352.

Leopold, L. B. 1977. A Reference for Rivers. *Geology* 5:429–430.

Leopold, L. B., and T. Maddock, Jr. 1953. The Hydraulic Geometry of Stream Channels and Some Physiographic Implications. *U.S. Geological Survey Professional Paper 252.* 57 pp. Washington, D.C.: GPO.

Leopold, L. B., and M. G. Wolman. 1957. River Channel Patterns: Braided, Meandering, and Straight. *U.S. Geological Survey Professional Paper 282-B,* 39–73. Washington, D.C.: GPO.

Leopold, L. B., M. G. Wolman, and J. Miller. 1964. *Fluvial Processes in Geomorphology.* San Francisco, Calif.: W. H. Freeman.

Lev, E. 1995. Youth Conservation Corps Carries Out Streambank Project. *Restoration and Management Notes* 13:20–21.

Lewin, J. 1978. Floodplain Geomorphology. *Progressive Physical Geography* 2:408–437.

Liddle, M. J. 1975. A Selective Review of the Ecological Effects of Human Trampling on Natural Ecosystems. *Biological Conservation* 7:17–36.

Little, W. C., C. R. Thorne, and J. B. Murphey. 1982. Mass Bank Failure Analysis of Selected Yazoo Basin Streams, Mississippi River Basin. *Transactions of the American Society of Agricultural Engineers* 25:1321–1328.

Love, D. W. 1979. Quaternary Fluvial Geomorphic Adjustments in Chaco Canyon. In *Adjustment of the Fluvial System,* edited by D. D. Rhodes and G. P. Williams, 277–280. Dubuque, Iowa: Kendall-Hunt.

Lutz, H. 1945. Soil Conditions on Picnic Grounds in Public Forest Parks. *Journal of Forestry* 43:121–127.

Mackin, J. H. 1948. *Concept of the Graded River.* Geological Society of America Bulletin, vol. 59.

Maddock, T., Jr. 1976. A Primer on Floodplain Dynamics. *Journal of Soil and Water Conservation* 31:44–47.

Magil, A. W., and A. T. Leiser. 1967. *New Help for Worn Out Campgrounds.* Berkeley, Calif.: USDA Pacific Southwest Forest and Range Experiment Station.

Magil, A. W., and R. H. Twiss. 1965. *A Guide for Recording Esthetic and Biologic Changes with Photographs*. USDA Forest Service Research Notes PSW-77. Washington, D.C.: U.S. Department of Agriculture.

Manci, K. M. 1989. Riparian Ecosystem Creation and Restoration: A Literature Summary. *U.S. Fish and Wildlife Service Biological Report* 89:59.

Manning, R. E. 1979. Impacts of Recreation on Riparian Soil and Vegetation. *Water Resources Bulletin* 15:30–43.

Marks, J. B. 1950. Vegetation and Soil Relations in the Lower Colorado Desert. *Ecology* 31:176–193.

McBride, J. R., and J. Strahan. 1984. Establishment and Survival of Woody Riparian Species on Gravel Bars of an Intermittent Stream. *American Midland Naturalist* 112:235–245.

McNatt, R. M., R. J. Hallock, and A. W. Anderson. 1980. *Riparian Habitat and Instream Flow Studies: Lower Verde River*. Albuquerque, N.Mex.: Riparian Habitat Analysis Group, Office of Environment, U.S. Fish and Wildlife Service, Region 2.

McNeal, B. L., W. A. Norvell, and N. T. Coleman. 1966. Effect of Solution Composition on the Swelling of Extracted Soil Clays. *Proceedings of the Soil Science Society of America* 30:313–317.

Medina, A. L. 1990. Possible Effects of Residential Development on Streamflow, Riparian Plant Communities, and Fisheries on Small Mountain Streams in Central Arizona. *Forest Ecology and Management* 33/34:351–361.

Meehan, W. R., and W. S. Platts. 1978. Livestock Grazing and the Aquatic Environment. *Journal of Soil and Water Conservation* 33:274–278.

Merriam, L. C., Jr., and C. K. Smith. 1974. Visitor Impact on Newly Developed Campsites in the Boundary Waters Canoe Area. *Journal of Forestry* 72:627–630.

Miller, C. R., and W. M. Borland. 1963. Stabilization of Fivemile and Muddy Creeks. *American Society of Civil Engineers, Journal of Hydraulics Division* 89:67–98.

Minckley, W. L., and T. O. Clark. 1984. Formation and Destruction of a Gila River Mesquite Bosque Community. *Desert Plants* 6:23–30.

Minnich, D. W. 1978. An Overview of the Demands upon Lowland River and Stream Riparian Habitat in Colorado. In *Lowland River and Stream Habitat in Colorado: A Symposium (Oct. 4–5, 1978, Greeley, Colorado)*, coordinated by W. D. Graul and S. J. Bissell, 1–42. Greeley: University of Northern Colorado.

Moir, W. H. 1989. History of Development of Site and Condition Criteria for Range Condition within the U.S. Forest Service. In *Secondary Succession and Evaluation of Rangeland Condition*, edited by W. K. Lauenroth and W. A. Laycock, 49–76. Boulder, Colo.: Westview Press.

Mollard, J. D. 1973. Airphoto Interpretation of Fluvial Features. In *Proceedings of the 9th Canadian Hydrology Symposium, Fluvial Processes and Sedimentation*, Research Council of Canada, 341–380. Ottawa: Government Printing.

Mooney, H. A., B. B. Simpson, and O. T. Solbrig. 1977. Phenology, Morphology, Physiology. In *Mesquite: Its Biology in Two Desert Ecosystems*, edited by B. B. Simpson, 26–43. Stroudsburg, Penn.: Dowden, Hutchinson and Ross.

Moore, S. 1989. Visitor Management in a Riparian Wilderness Habitat. Paper presented at the Fourth Annual Conference of the Riparian Council, September 22–23, Sun Rise Ski Resort, McNary, Arizona.

Moreno, J. L. 1992. Probables impactos ambientales del tratado de libro comercio en el estado de Sonora. *Ecologia, recursos naturales y medio ambiente en Sonora*, 349–360.

Morisawa, M. E. 1985. *Rivers, Form, and Process.* Geomorphology Texts 7. London: Longman.

Murdoch, T. 1995. Stream Restoration and Environmental Education: The Adopt-A-Stream Foundation. *Restoration and Management Notes* 13:7–11.

Neill, W. M. 1990. Control of Tamarisk by Cut-Stump Herbicide Treatments. In *Tamarisk Control in Southwestern United States,* 91–98. Cooperative National Park Resources Studies Unit, Special Report no. 9. Tucson: School of Renewable Natural Resources, University of Arizona.

Nixon, E. S., R. L. Wiollet, and P. W. Cox. 1977. Woody Vegetation in a Virgin Forest in an Eastern Texas River Bottomland. *Castanea* 42:227–236.

Ohmart, R. D., and B. W. Anderson. 1986. Riparian Habitat. In *Inventory and Monitoring of Wildlife Habitat,* edited by A. Y. Cooperrider, R. J. Boyd, and H. R. Stuart, 169–199. Denver: USDI Bureau of Land Management Service Center.

Ohmart, R. D., W. O. Deason, and C. Burke. 1977. A Riparian Case History: The Colorado River. In *Importance, Preservation, and Management of Riparian Habitat: A Symposium (July 9, 1977, Tucson, Arizona),* coordinated by R. R. Johnson and D. A. Jones, 35–47. USDA Forest Service General Technical Report RM-43. Fort Collins, Colo.: Rocky Mountain Forest and Range Experiment Station.

Ohmart, R. D., and C. D. Zisner. 1993. *Functions and Values of Riparian Habitat to Wildlife in Arizona: A Literature Review.* Report to the Arizona Game and Fish Department, Phoenix, Contract no. G30025-B.

Oster, J. D., and L. S. Willardson. 1971. Reliability of Salinity Sensors for the Measurement of Soil Salinity. *Agronomy Journal* 63:695–698.

Osterkamp, W. R. 1978. Gradient, Discharge, and Particle-Size Relations of Alluvial Channels in Kansas, with Observations on Braiding. *American Journal of Science* 278:1253–1268.

Osterkamp, W. R. 1979. Bed- and Bank Material Sampling Procedures at Channel-Geometry Sites. In *Proceedings, National Conference on Quality Assurance of Environmental Measurements (November 27–29, 1978, Denver, Colo.),* 86–89. Silver Springs, Md.: Information Transfer.

Osterkamp, W. R., and P. E. Harrold. 1982. Dynamics of Alluvial Channels — A Process Model. In *Modeling Components of the Hydrologic Cycle: Proceedings, International Symposium on Rainfall-Runoff Modeling,* 283–296. Littleton, Colo.: Water Resources Publications.

Osterkamp, W. R., and E. R. Hedman. 1977. Variation of Width and Discharge for Natural High-Gradient Channels. *Water Resources Research* 13(2):256–258.

Osterkamp, W. R., and C. R. Hupp. 1984. Geomorphic and Vegetative Characteristics along Three Northern Virginia Streams. *Geological Society of America Bulletin* 5:1093–1101.

Patterson, D. W., C. U. Finch, and G. I. Wilcox. 1981. Streambank Stabilization Techniques Used by the Soil Conservation Service. In *California Riparian Systems: Ecology, Conservation, and Productive Management,* edited by R. E. Warner and K. M. Hendrix, 452–459. Berkeley and Los Angeles: University of California Press.

Pendleton, D. T. 1989. Range Condition as Used in the Soil Conservation Service. In *Secondary Succession and Evaluation of Rangeland Condition,* edited by W. K. Lauenroth and W. A. Laycock, 17–34. Boulder, Colo.: Westview Press.

Petts, G. E., and I. Foster. 1985. *Rivers and Landscape.* London: Edward Arnold Publishers.

Pinkney, F. C. 1992. *Revegetation and Enhancement of Riparian Communities along the Lower Colorado River.* Denver: USDI Bureau of Reclamation, Ecological Resources Branch.

Platts, W. S. 1989. Compatibility of Livestock Grazing Strategies with Fisheries. In *Practical Approaches to Riparian Resource Management: An Educational Workshop,* 103–110. Bethesda, Md.: American Fisheries Society.

Platts, W. S., and R. L. Nelson. 1985a. Impacts of Rest-Rotation Grazing on Stream Banks in Forested Watersheds in Idaho. *North American Journal of Fisheries Management* 5:547–556.

Platts, W. S., and R. L. Nelson. 1985b. Will the Riparian Pasture Build Good Streams? *Rangelands* 7:7–10.

Platts, W. S., and R. L. Nelson. 1989. Characteristics of Riparian Plant Communities and Streambanks with Respect to Grazing in Northeastern Utah. In *Practical Approaches to Riparian Resource Management: An Educational Workshop (May 8–11, Billings, Montana),* edited by R. E. Gresswell, B. A. Barton, and J. L. Kershner, 73–81. Billings, Mont.: U.S. Bureau of Land Management.

Platts, W. S., and R. F. Raleigh. 1984. Impacts of Grazing on Wetlands and Riparian Habitat. In *Developing Strategies for Rangeland Management,* National Research Council\ National Academy of Sciences, 1105–1117. Boulder, Colo.: Westview Press.

Platts, W. S., and F. J. Wagstaff. 1984. Fencing to Control Livestock Grazing on Riparian Habitats along Streams: Is It a Viable Alternative? *North American Journal of Fisheries Management* 4:266–272.

Platts, W. S., F. J. Wagstaff, and E. Chaney. 1989. Cattle and Fish on the Henry's Fork. *Rangelands* 11:58–62.

Porter, H. L., and L. F. Silberberger. 1961. Streambank Stabilization. *Journal of Soil and Water Conservation* 16:214–216.

Post, D. F. 1979. *Soil Conditions on Campsite and Recreational Areas with Special Reference to Bonita Campground, Chiricahua National Monument, in Arizona.* Cooperative National Park Resources Studies Unit, Technical Report no. 4. Tucson: School of Renewable Natural Resources, University of Arizona.

Reichard, N. 1984. Riparian Habitat Restoration: Some Techniques for Dealing with Landowners, Livestock, and Eroding Streambanks. In *Proceedings of Pacific Northwest Stream Habitat Management Workshop (October 10–12, 1984, Arcata, Calif.),* edited by T. J. Hassler. Arcata, Calif.: Humboldt State University.

Reichenbacher, F. W. 1984. Ecology and Evaluation of Southwestern Riparian Plant Communities. *Desert Plants* 6:15–22.

Renner, F. G., and B. W. Allred. 1962. *Classifying Rangeland for Conservation Planning.* U.S. Department of Agriculture Handbook 253. Washington, D.C.: GPO.

Rhoades, J. D. 1976. Measuring, Mapping, and Monitoring Field Salinity and Water Table Depths with Soil Resistance Measurements. *FAO Soils Bulletin* 31:159–186.

Rhoades, J. D., P.A.C. Raats, and R. J. Prather. 1976. Effects of Liquid-Phase Electrical Conductivity on Bulk Soil Electrical Conductivity. *Soil Science Society of America Journal* 43:817–818.

Rhoads, B. L., and M. V. Miller. 1990. Impact of Riverine Wetlands Construction and Operation on Stream Channel Stability: Conceptual Framework for Geomorphic Assessment. *Environmental Management* 14(6):799–807.

Richter, H. 1992. Development of a Conceptual Model for Floodplain Restoration. *Arid Lands Newsletter* 32:13–17.

Robinson, T. W. 1965. Introduction, Spread, and Areal Extent of Saltcedar *(Tamarix)* in Western States. *U.S. Geological Survey Professional Paper 491-A.* 12 pp. Washington, D.C.: GPO.

Roundy, B. A., R. A. Evans, and J. A. Young. 1984. Surface Soil and Seedbed Ecology in Salt-Desert Plant Communities. In *Proceedings — Symposium on the Biology of Atriplex and Related Chenopods (May 2–6, 1983, Provo, Utah),* compiled by A. R. Tiedemann, E. D. McArthur, H. C. Stutz, R. Stevens, and K. L. Johnson, 66–74. USDA Forest Service General Technical Report INT-172. Ogden, Utah: Intermountain Forest and Range Experiment Station.

Rzhanitsyn, N. A. 1960. *Morphological and Hydrological Regularities of the Structure of the River Net.* Translated by D. B. Krimgold. U.S. Department of Water Research Division and U.S. Department of the Interior, Geological Survey, Water Resources Division. Washington, D.C.: GPO.

Sanchez, S. P. 1991. Impacto de las actividades primarias sobre los ecosistemas en el estado de Sonora. *Ecologia y medio ambiente,* 4761.

Schmidt, L. J. 1987. Recognizing and Improving Riparian Values: The Forest Service Approach to Riparian Management. In *Proceedings of the Society of Wetland Scientists' Eighth Annual Meeting (May 26–29, Seattle, Wash.),* edited by K. M. Mutz and L. C. Lee, 36–39. Denver: Planning Information Corporation.

Schultze, R. F., and G. I. Wilcox. 1985. Emergency measures for streambank stabilization: An evaluation. In *Riparian Ecosystems and Their Management: Reconciling Conflicting Uses (First North American Riparian Conference, April 16–18, 1985, Tucson, Arizona),* coordinated by R. R. Johnson, C. D. Ziebell, D. R. Patton, P. F. Ffolliot, and R. H. Hamre, 59–61. USDA Forest Service General Technical Report RM-120. Fort Collins, Colo.: Rocky Mountain Forest and Range Experiment Station.

Schumm, S. A. 1977. *The Fluvial System.* New York: John Wiley.

Schumm, S. A. 1981. Evolution and Response of the Fluvial System: Sedimentological Implications. *Society of Economic Paleontologists and Mineralogists Special Publication* 31:19–29.

Schumm, S. A., M. D. Harvey, and C. C. Watson. 1984. *Incised Channels Morphology, Dynamics, and Control.* Littleton, Colo.: Water Resources Publications.

Settergren, C. D. 1977. Impacts of River Recreation Use on Streambank Soils and Vegetation: State-of-the-Knowledge. In *River Recreation Management and Research Symposium,* 55–59. USDA Forest Service General Technical Report NC-28. St. Paul, Minn.: North Central Forest Experimental Station.

Settergren, C. D., and D. M. Cole. 1970. Recreation Effects on Soil and Vegetation in the Missouri Ozarks. *Journal of Forestry* 68:231–234.

Shen, H. W., and S. Vedula. 1969. A Basic Cause of a Braided Channel. In *Proceedings of the 13th Congress of the International Association for Hydraulic Research,* 201–205. Kyoto, Japan: International Association for Hydraulic Research.

Siegel, R. S., and J. H. Brock. 1990. Germination Requirements of Key Southwestern Woody Riparian Species. *Desert Plants* 10:3–8.

Simcox, D. E., and E. H. Zube. 1985. *Arizona Riparian Areas: A Bibliography.* Prepared for the First North American Riparian Conference — Riparian Ecosystems and Their Management: Reconciling Conflicting Uses, April 16–18, 1985, Tucson, Arizona. Tucson: School of Renewable Natural Resources, University of Arizona.

Simons, D. B., and E. V. Richardson. 1966. Resistance to Flow in Alluvial Channels. *U.S. Geological Survey Professional Paper 422-J.* Washington, D.C.: GPO.

Simpson, B. B. 1977. *Mesquite: Its Biology in Two Desert Scrub Ecosystems.* Stroudsburg, Penn.: Dowden, Hutchinson and Ross.

Singer, M. J., and D. N. Munns. 1987. *Soils.* New York: Macmillan.

Skoglund, J. 1990. Seed Dispersing Agents in Two Regularly Flooded River Sites. *Canadian Journal of Botany* 68:745–760.

Skovlin, J. M. 1984. Impacts of Grazing on Wetlands and Riparian Habitat: A Review of Our Knowledge. In *Developing Strategies for Rangeland Management: A Report Prepared by the Committee on Developing Strategies for Rangeland Management,* National Research Council/National Academy of Sciences, 1001–1103. Boulder, Colo.: Westview Press.

Smith, S. D., A. B. Wellington, J. L. Nachlinger, and C. A. Fox. 1991. Functional Responses of Riparian Vegetation to Streamflow Diversion in the Eastern Sierra Nevada. *Ecological Applications* 1:89–97.

Snyder, C. T., D. G. Frickel, R. F. Hadley, and R. G. Miller. 1976. *Effects of Off-Road Vehicle Use on the Hydrology and Landscape of Arid Environments in Central and Southern California.* U.S. Geological Survey Water Resources Investigations no. 4. Denver.

Spear, M. J., and C. L. Mullins. 1987. Riparian Habitat of the Middle Rio Grande: A Case Study for More Effective Protection. In *Proceedings of the Society of Wetland Scientists' Eighth Annual Meeting (May 26–29, Seattle, Wash.),* edited by K. M. Mutz and L. C. Lee, 45–48. Denver: Planning Information Corporation.

Stoddart, L. A., A. D. Smith, and T. W. Box. 1975. *Range Management.* 3rd ed. New York: McGraw-Hill.

Strahan, J. 1981. Regeneration of Riparian Forest of the Central Valley. Paper presented at the California Riparian Systems Conference, September 17–19, University of California, Davis.

Streng, D. R., J. S. Glitzenstein, and P. A. Harcombe. 1989. Woody Seedling Dynamics in an East Texas Floodplain Forest. *Ecological Monographs* 59:177–204.

Stromberg, J. C. 1992. Riparian Mesquite Forests: A Review of Their Ecology, Threats, and Recovery Potential. *Journal of the Arizona-Nevada Academy of Science* 27:111–124.

Stromberg, J. C. 1993. Fremont Cottonwood–Goodding Willow Riparian Forests: A Review of Their Ecology, Threats, and Recovery Potential. *Journal of the Arizona-Nevada Academy of Science* 26:97–100.

Stromberg, J. C., D. T. Patten, and B. D. Richter. 1991. Flood Flows and Dynamics of Sonoran Riparian Forests. *Rivers* 2:221–235.

Stromberg, J. C., M. R. Sommerfeld, D. T. Patten, J. Fry, C. Kramer, F. Amalfi, and C. Christian. 1993. Release of Effluent into the Upper Santa Cruz River: Ecological Considerations. In *Proceedings of the Symposium on Effluent Use Management and Abstracts, AWRA 19th Annual Conference (August 29–September 2, Tucson, Ariz.),* edited by K. D. Schmidt and M. G. Wallace, 81–90. Bethesda, Md.: American Water Resources Association.

Stromberg, J. C., J. A. Tress, S. D. Wilkins, and S. D. Clark. 1992. Response of Velvet Mesquite to Groundwater Decline. *Journal of Arid Environments* 23:45–58.

Sudbrock, A. 1993. Tamarisk Control 1: Fighting Back — An Overview of the Invasion, and a Low-Impact Way of Fighting It. *Restoration and Management Notes* 11:31–34.

Suryanarayana, B. 1969. Mechanics of Degradation and Aggradation in a Laboratory Flume. Ph.D. diss., Colorado State University.

Svejcar, T. 1989. Streambank Plants Vital to Water Quality. *Agricultural Research* 8:19.

Swanson, S. 1988. Riparian Values as a Focus for Range Management and Vegetation

Science. In *Vegetation Science Applications for Rangeland Analysis and Management,* edited by P. T. Tueller, 105–126. Boston, Mass.: Kluwer.

Swanson, S., S. Franzen, and M. Manning. 1987. Rodero Creek: Rising Water on the High Desert. *Journal of Soil and Water Conservation* 42:405–407.

Swenson, E. A., and C. L. Mullins. 1985. Revegetating Riparian Trees in Southwestern Floodplains. In *Riparian Ecosystems and Their Management: Reconciling Conflicting Uses (First North American Riparian Conference, April 16–18, 1985, Tucson, Arizona),* coordinated by R. R. Johnson, C. D. Ziebell, D. R. Patton, P. F. Ffolliot, and R. H. Hamre, 135–139. USDA Forest Service General Technical Report RM-120. Fort Collins, Colo.: Rocky Mountain Forest and Range Experiment Station.

Szaro, R. C. 1989. Riparian Forest and Scrubland Community Types of Arizona and New Mexico. *Desert Plants* 9:1–138.

Szaro, R. C. 1990. Southwestern Riparian Plant Communities: Site Characteristics, Tree Species Distributions, and Size-Class Structures. *Forest Ecology Management* 33/34: 315–334.

Taub, F. B. 1987. Indicators of Change in Natural and Human-Impacted Ecosystems: Status. In *Preserving Ecological Systems: The Agenda for Long-Term Research and Development,* edited by S. Draggan, J. J. Cohrssen, and R. E. Morrison, 115–144. New York: Praeger.

Tellman, B., and R. Jemison. 1995. *Riparian Wetland Research Expertise Directory: Arizona, Colorado, Nevada, New Mexico, and Utah.* Tucson, Ariz., and Fort Collins, Colo.: joint publication by the Water Resources Research Center of the University of Arizona and the USDA Forest Service's Rocky Mountain Forest and Range Experiment Station.

Teskey, R. O., and T. M. Hinckley. 1977. *Impact of Water Level Changes on Woody Riparian and Wetland Communities,* vol. 1, *Plant and Soil Responses to Flooding.* Columbia, Mo.: National Stream Alteration Team, Office of Biological Services, U.S. Fish and Wildlife Service.

Thorne, C. R. 1981. Field Measurements of Rates of Bank Erosion and Bank Material Strength, Soil Erodibility, Shear Strength, Tensile Strength. In *Erosion and Sediment Transport Measurement Symposium: Proceedings (June 22–26, Florence, Italy).* IAHS-AISH Publications no. 133. Washington, D.C.: International Association of Hydrological Sciences.

Todd, D. K. 1980. *Groundwater Hydrology.* New York: John Wiley & Sons.

Toy, T. J., and R. F. Hadley. 1987. *Geomorphology and Reclamation of Disturbed Lands.* Orlando, Fla.: Academic Press and Harcourt Brace Jovanovich.

Turner, R. M. 1974. Quantitative and Historical Evidence of Vegetation Changes along the Upper Gila River, Arizona. *U.S. Geological Survey Professional Paper 655-H,* 1–20. Washington, D.C.: GPO.

U.S. Salinity Laboratory Staff. 1954. *Diagnosis and Improvement of Saline and Alkaline Soils.* USDA Handbook no. 60. Washington, D.C.: GPO.

Van der Moezel, P. G., L. E. Watson, G.V.N. Pearce-Pinto, and D. T. Bell. 1988. The Response of Six Eucalyptus Species and *Casuarina obesa* to the Combined Effect of Salinity and Waterlogging. *Australian Journal of Plant Physiology* 15:44–67.

Van Haveren, B. P., and W. L. Jackson. 1986. Concepts in Stream Riparian Rehabilitation. In *Proceedings of the Wildlife Management Institute's 51st North American Wildlife and Natural Resources Conference (March 21–26, Reno, Nevada),* 1–18. Washington, D.C.: Wildlife Management Institute.

Verstappen, H. Th. 1983. *Applied Geomorphology: Geomorphologic Surveys for Environmental Development.* Amsterdam: Elsevier.

Wager, J. A. 1966. Cultural Treatment of Vegetation on Recreation Sites. In *Proceedings of the Society of American Foresters,* 37–39. Washington D.C.: Society of American Foresters.

Wahl, K. L. 1977. Accuracy of Channel Measurements and Their Implications in Estimating Stream Flow Characteristics. *Journal of Research, U.S. Geological Survey* 5:811–814.

Wallace, D. E., and L. J. Lane. 1976. Geomorphic Thresholds and Their Influence on Surface Runoff from Small Semiarid Watersheds. *Hydrology and Water Resources in Arizona and the Southwest* 6:169–176 [Office of Arid Land Studies, University of Arizona, Tucson].

Walling, D. E. 1978. Reliability Considerations in the Evaluation and Analysis of River Loads. *Zeitschrift für Geomorphologie* Supplement NF 29:29–42.

Walters, M. A., R. O. Teskey, and T. M. Hinckley. 1980. *Impact of Water Level Changes on Woody Riparian and Wetland Communities,* vol. 7. Kearneysville, W.Va.: Eastern Energy and Land Use Team, National Water Resources Analysis Group, Office of Biological Services, U.S. Fish and Wildlife Service.

Warner, R. E., and K. M. Hendrix. 1984. *California Riparian Systems: Ecology, Conservation, and Productive Management.* Berkeley: University of California Press.

Warren, D. K., and R. M. Turner. 1975. Saltcedar *(Tamarix chinensis)* Seed Production, Seedling Establishment, and Response to Inundation. *Journal of the Arizona Academy of Science* 10:135–144.

Warren, P. L., and S. L. Anderson. 1985. Gradient Analysis of a Sonoran Desert Wash. In *Riparian Ecosystems and Their Management: Reconciling Conflicting Uses (First North American Riparian Conference, April 16–18, 1985, Tucson, Arizona),* coordinated by R. R. Johnson, C. D. Ziebell, D. R. Patton, P. F. Ffolliot, and R. H. Hamre, 150–155. USDA Forest Service General Technical Report RM-120. Fort Collins, Colo.: Rocky Mountain Forest and Range Experiment Station.

Way, D. S. 1978. *Terrain Analysis: A Guide to Site Selection Using Aerial Photographic Interpretation.* Stroudsburg, Penn.: Dowden, Hutchinson and Ross.

Webb, R. H. 1983. Compaction of Desert Soils by Off-Road Vehicles. In *Environmental Effects of Off-Road Vehicles,* edited by R. H. Webb and H. G. Wilshire, 51–79. New York: Springer-Verlag.

White, P. S. 1979. Pattern, Process, and Natural Disturbance in Vegetation. *Botanical Review* 45:229–299.

Willard, B. E., and J. W. Marr. 1970. Effects of Human Activities on Alpine Tundra Ecosystems in Rocky Mountain National Park, Colorado. *Biological Conservation* 2:257–265.

Wolman, M. G. 1955. The Natural Channel of Brandywine Creek, Pennsylvania. *U.S. Geological Survey Professional Paper 271,* 1–18. Washington, D.C.: GPO.

Wolman, M. G. 1967. Two Problems Involving River Channel Changes and Background Observations. *Northwestern University Studies in Geography* 4:67–107.

Wolman, M. G., and L. B. Leopold. 1957. River Flood Plains: Some Observations on Their Formation. *U.S. Geological Survey Professional Paper 282-C,* 87–107. Washington, D.C.: GPO.

Wolt, J. 1994. *Soil Solution Chemistry: Applications to Environmental Science and Agriculture.* New York: John Wiley & Sons.

Yadav, B. R., N. H. Rao, K. V. Paliwal, and P.B.S. Sarma. 1979. Comparison of Different Methods for Measuring Soil Salinity under Field Conditions. *Soil Science* 127:335–339.

Yair, A., D. Sharon, and H. Lavee. 1978. An Instrumented Watershed for the Study of Partial Area Contribution of Runoff in the Arid Zone. *Zeitschrift für Geomorphologie* Supplement NF 29:71–82.

Young, R. A., and A. R. Gilmore. 1976. Effects of Various Camping Intensities on Soil Properties in Illinois Campgrounds. *Soil Science Society of America Journal* 40:908–911.

Young, J. A., and C. G. Young. 1986. *Collecting, Processing, and Germinating Seeds of Wildland Plants*. Portland, Oreg.: Timber Press.

Young, J. A., and C. G. Young. 1992. *Seeds of Woody Plants in North America*. Portland, Oreg.: Dioscorides Press.

Youngblood, A. P., W. G. Padgett, and A. H. Winward. 1985. Riparian Community Type Classification in the Intermountain Region. In *Riparian Ecosystems and Their Management: Reconciling Conflicting Uses (First North American Riparian Conference, April 16–18, 1985, Tucson, Arizona)*, coordinated by R. R. Johnson, C. D. Ziebell, D. R. Patton, P. F. Ffolliot, and R. H. Hamre, 510–512. USDA Forest Service General Technical Report RM-120. Fort Collins, Colo.: Rocky Mountain Forest and Range Experiment Station.

Zimmerman, R. C. 1969. Plant Ecology of an Arid Basin, Tres Alamos–Redington Area, Southeastern Arizona. *U.S. Geological Survey Professional Paper 485-D*. 171 pp. Washington, D.C.: GPO.

Personal Communications

Anderson, Bertin. 1990, 1992, 1993, and 1994. Revegetation and Wildlife Management Center, Blythe, Ariz.

Bammann, Al. 1990. U.S. Bureau of Land Management, Safford Resource Area, Safford, Ariz.

Bell, Gary. 1990. U.S. Forest Service, Sitgreaves National Forest, Pleasant Valley Ranger District, Young, Ariz.

Castellanos, Alejandro. 1994. Jefe de Departamento, Centro de Investigaciones Científicas y Tecnológicas de la Universidad de Sonora, Hermosillo, Sonora, Mexico.

Fenner, Pattie. 1995. U.S. Forest Service, Cave Creek Ranger District, Watershed Division, Carefree, Ariz.

Forbis, Larry. 1990. U.S. Forest Service, Tonto National Forest, Mesa Ranger District, Mesa, Ariz.

Francois, L. E. 1992 and 1993. U.S. Salinity Laboratory, Riverside, Calif.

Goodwin, Greg. 1990. U.S. Forest Service, Coconino National Forest, Supervisor's Office, Flagstaff, Ariz.

Hall, Robert. 1990, 1992. U.S. Bureau of Land Management, Kingman Resource Area, Kingman, Ariz.

Moore, Catsby. 1990. Maricopa County Flood Control, Phoenix.

Pollock, Don. 1990. U.S. Forest Service, Payson Ranger District, Payson, Ariz.

Richter, Brian and Holly Richter. 1990. Nature Conservancy, Hassayampa River Preserve, Wickenburg, Ariz.

Robles, Ben. 1995. U.S. Bureau of Land Management, Safford, Ariz.

Scott, Michael. 1995. U.S. Fish and Wildlife Service, Fort Collins, Colo.

Sims, William. 1990. U.S. Bureau of Indian Affairs, San Carlos Agency, San Carlos, Ariz.

York, John. 1990. U.S. Soil Conservation Service, Phoenix.

Zarlingo, Vern. 1990. U.S. Forest Service, Kaibab National Forest, Williams, Ariz.

Index

About the Author

Mark K. Briggs is director of research at the Rincon Institute in Tucson, Arizona. Briggs studied at the School of Renewable Natural Resources of the University of Arizona. He has extensive experience in evaluating riparian restoration efforts in the southwestern United States and has enjoyed several long stints in West Africa. Briggs conducts field workshops on evaluating and restoring degraded riparian areas and has developed recovery plans for the lower Gila River, Little Colorado River, and streams near Tucson, Arizona. He has also developed a long-term riparian monitoring program that will improve the state of knowledge of how urbanization and other human disturbances influence the condition of riparian ecosystems. Briggs was recently appointed to the City of Tucson's Urban Runoff Technical Committee to advise the city on ways to manage runoff while preserving streamside vegetation communities.